W9-BCN-391

Our Changing Planet

85200

Our Changing Planet

John Gribbin

550
G 871o

Thomas Y. Crowell Company
Established 1834
New York

MERNER - PFEIFFER LIBRARY
TENNESSEE WESLEYAN COLLEGE
ATHENS, TN. 37303

Copyright ©1977 by John Gribbin

All rights reserved. Except for use in a review, the reproduction or utilization of this work in any form or by any electronic, mechanical, or other means, now known or hereafter invented, including xerography, photocopying, and recording, and in any information storage and retrieval system is forbidden without the written permission of the publisher.

Manufactured in the United States of America

Library of Congress Cataloging in Publication Data

Gribbin, John R
Our changing planet.

Bibliography: p.
Includes index.
1. Geology. I. Title
QE31.G82 550 77–5547
ISBN 0–690–01693–X

OCT 30 '78

Contents

For BENJAMIN
who tries harder

Introduction

In recent years much has been made of the concept of 'Planet Earth', with its limited resources and thin layer of atmosphere and ocean – the biosphere. We have been warned that the finite extent of the biosphere makes it susceptible to man-made pollution, and that the finite resources of Planet Earth could be exhausted by indiscriminate consumption in our modern industrialized society. Either eventuality could spell the end of that society. But man and all his works are of no great importance to Planet Earth, which has experienced dramatic changes over hundreds of millions of years. The forces which have operated to produce the world we know today are still at work, changing both the face of the Earth and the nature of the biosphere on which we depend for life.

Creatures like ourselves need air and water, winds, oceans and land; we have evolved to suit a particular arrangement of these physical properties of our planet. But these features are not permanent, and contain within themselves the seeds of their own destruction or alteration. Earthquakes, ice ages, the advance and retreat of deserts, mountain-building and the development of oceans are all natural features of our changing planet. And all involve potentially disastrous situations for mankind, far more dangerous than any tinkering we may do to Planet Earth.

However, there is another side to the coin. With a proper understanding of the underlying causes of these changes, we can gain a better insight into the workings of our planet. We can decide which regions of the world, for example, are likely to have as yet untapped reserves of oil, coal or metal ore, and by understanding the processes which bring about earthquakes, say, we can learn how to avoid their most damaging effects.

This kind of understanding has become possible in recent years, mainly through the development of the modern theory of plate tectonics, which explains how the continents of the Earth move about on the surface of the globe. But these geophysical ideas cannot stand alone; to gain a full insight into what goes on around us they must be combined with ideas from other specialist disciplines – meteorology and oceanography, for instance. The best picture of the true significance (or, rather, insignificance) of our home in space comes, however, from a scientific discipline not usually regarded as applying to the Earth – astronomy. Astronomers have long been aware of the finite extent of our home in both space and time, and their studies of planets in general can be applied fruitfully to the special case of the planet on which we live. The story of the changing Earth began more than 5,000 million years ago, with the formation of the Solar System from a cloud of dust and gas in interstellar space. Modern astronomical ideas give us a very good concept of what things must have been like in the beginning, and an outline, at least, of how our planet has developed since then.

In a very real sense, then, the more detailed study of our own planet and its development – geophysics – is a special branch of planetary astronomy, and geophysicists can learn a great deal about the nature of Planet Earth by making comparisons with the evidence now available from unmanned spacecraft sent to visit other planets (equally, of course, planetary astronomers can interpret the evidence from, say, Mars much more satisfactorily since they have a reasonable understanding of the Earth already). But whether you label the story of the changing Earth 'geophysics' or 'astronomy' the story remains the same, and it is of vital importance today because for the first time the new Earth science disciplines can not only tell us that our resources are limited, but can also provide an indication of just how limited they are. It is the difference between driving along an intercontinental highway at full speed, knowing vaguely that the road must stop somewhere, and seeing the first warning signs indicating how many hundred metres remain before the turn-off. As all drivers know, the sooner these warning signs are seen the better; in this book I have attempted to combine both warning signs and map, showing just how far along its road the Earth has already travelled, examining where it is now, and sketching in something of the path ahead.

Chapter One
In the Beginning

Our Earth is one of nine planets orbiting an unexceptional star which in turn orbits around the centre of a large aggregation of stars, our Galaxy. The Galaxy itself is nothing special on the astronomical scale of things, and there are millions of other galaxies in the Universe. However, it has the greatest significance for us and our Solar System, because it is our home. And it now seems that the structure of the Galaxy has had a great and continuing significance for the formation of our Solar System and for events that have occurred on Earth over the past few hundred million years.

Our own Milky Way Galaxy contains about a trillion stars, most of them concentrated in a flat disk about 100,000 light years across and some 2,000 light years thick (one light year is the distance covered by light in a year, travelling at a speed of 30,000 million centimetres every *second*). This pancake of stars forms a spiral pattern, rather like the galaxy M51 shown in Figure 1, and like many other galaxies. Individual stars orbit around the centre of the Galaxy under the influence of gravity; but the spiral pattern itself seems to be more or less fixed, and this has vital consequences for the existence of our Solar System – and of the Earth – in the form we see them.

Ten thousand million years ago, there were probably no stars at all in the Galaxy – it just consisted of a cloud of gas in space. Since then, parts of that original cloud have condensed to form stars, in groups ranging from a few stars to great clusters containing several million. But the early history of the Galaxy need not concern us; our Solar System did not form until the spiral pattern was well established, and the Galaxy contained both stars and clouds of gas and dust, all orbiting around the centre and passing repeatedly through the spiral arms. This repeated passage through the arms seems to be the key to the existence of our Earth.

The speed of the Solar System today carries it around the Galaxy in one orbit every 250 million years. This sounds a long time, but it means that our Sun and its family of planets have been round the Galaxy twenty times since they formed 5,000 million years ago. And before that, the cloud from which the Solar System formed must have been following the same orbit for hundreds of millions of years.

Like the Galaxy M51 shown in Figure 1, most spiral galaxies have two clearly defined spiral arms which twine around the galactic disk. Our own Galaxy is much the same as M51. The spiral arms themselves show up brightly because they contain many young, hot and bright stars; just along the inside edge of the arms, however, there are dark lanes of gas and dust. These dark lanes are the real heart of the spiral pattern.

Fig.1 Spiral Galaxy M51. (Hale Observatories)

According to modern theories, the spiral pattern is a standing shock wave. How it originates is still something of a mystery – but that mystery is outside the scope of this book. The dark lanes mark the boundary of the shock front. When a cloud of gas and dust, orbiting quietly around the centre of the Galaxy, arrives at one of these dark lanes it is subjected to violent forces which squeeze it into a smaller volume. For very diffuse clouds, this need not have any dramatic immediate consequences. But each time the cloud crosses a spiral arm it gets squeezed a stage further – and it crosses such an arm twice in every orbit of the Galaxy.

Eventually, something must happen to the cloud. Just what that something might be depends to a great extent on how dense the cloud was to start with, and how much matter there is in it. It could be squeezed just enough to produce a loose collection of material rather like a cloud of comets – indeed, this is one explanation of where comets come from. At the other extreme, a very large cloud compressed to the point where its own gravity causes it to break up and collapse completely would form a great cluster of stars. And somewhere in between these extremes we have a more ordinary cloud which, squeezed to the point of no return, collapses to produce a few stars and some miscellaneous rubble. These few stars soon separate from one another on their continuing journey around the Galaxy, each carrying with it some of the rubble, in the form of freshly created planets. Since the Solar System formed in this way, it has continued to travel around the Galaxy, crossing the spiral arms a further forty or fifty times. This has had a profound effect on events on Earth, as we shall see later. But for the time being, let's focus our attention within the Solar System, on the new planet Earth. How did it move from the primeval state to the present-day situation? And, indeed, how can we be so sure of the age of the Earth?

The second question, at least, is now fairly straightforward to answer. Many rocks contain, in small quantities, radioactive elements. These elements break down, through radioactive decay, into other, stable elements. And the rate at which this breakdown occurs is very strictly defined for each radioactive element. Take uranium, for example. A set quantity of uranium breaks down so that half of its mass (whatever amount you start with) is converted to other elements every 4,500 million years. In the case of uranium, the end product of this decay is lead; when a piece of rock forms, it has locked into it a fixed amount of uranium and other elements, and by measuring the proportions of uranium and lead in rocks today it is not difficult for geologists to work out both how old the rock is and how much uranium it contained originally.

Other radioactive elements are also involved in this technique, and the oldest rocks from the Earth's crust yet dated have ages of rather less than 4,000 million years. But Moon and meteorite rocks with ages greater than 4,500 million years have been identified, and the best estimate is that the age of the Solar System – and of the Earth – is something close to 5,000 million years.

To find out what the Earth was like just after it formed requires the combined efforts of geologists, geochemists, geophysicists, astronomers and biologists. The first thing that these different disciplines can tell us is that the Earth's present atmosphere is very different from any original atmosphere. The so-called noble gases (especially helium) are much less abundant on Earth than they are in the normal constituents of interstellar space, or the Sun and stars. It seems pretty clear that the processes associated with the formation of the Earth also resulted in these primeval gases being lost to space; then, a secondary atmosphere was formed out of chemical reactions from the rocks of the young planet.

When did the present atmosphere start to form? It is difficult to give a precise date, but the geological record shows that rocks originating by sedimentation were being laid down at least 3,000 million years ago; so there was an atmosphere and oceans (a hydrosphere) by that time. On the other hand, all the evidence implies that there was no free oxygen in the atmosphere at that time, nor was there for a further 1,200 million years or so. But then, 1,800 million years before the present, a dramatic change in the nature of the Earth's atmosphere occurred.

The early atmosphere seems to have been chiefly made up

of water, carbon dioxide and monoxide, nitrogen, hydrogen chloride, hydrogen and sulphur. These are just the gases which are released from volcanoes and hot springs, and it seems clear that they were liberated when the outer layers of our planet were heated by some dramatic event early in its life. One possibility is that the Moon was captured by the Earth about 3,500 million years ago, and that the resulting tidal friction caused rocks to melt and an atmosphere to form. There are other possibilities, but this is certainly a plausible and attractive idea.

Whatever the exact cause of the heating, an atmosphere did form. Microfossils are found in rocks more than 3,000 million years old, and these show that life originated long before there was any free oxygen in the atmosphere. Indeed, the oxygen only got into the atmosphere through the action of living organisms.

Thereby hangs a long and fascinating tale. In rocks between 1,800 and 3,200 million years old around the world there are many formations containing iron, which are known as the Banded Iron Formation, or B.I.F. It seems that early organisms (it hardly seems appropriate to call them primitive) made use of ferrous ions to absorb the oxygen which was being produced by photosynthesis. This oxygen would have been poisonous to all forms of life before the development of enzymes which modern life-forms use to mediate the activity of oxygen. So it was dumped, together with ferrous ions, in the form of the iron ores of the B.I.F.

Then, around 1,800 million years ago, the B.I.F. came to an end, at the same time as 'red beds' (material coated with ferric oxide) were first deposited. The conclusion seems inescapable; some organisms had now developed the necessary enzymes and had learnt to live with free oxygen. With no larger organisms to feed on them, these primitive algae must have soon dominated the oceans of the world, releasing great quantities of oxygen and thereby poisoning all the older forms of life, as well as causing the first-ever deposits of rust.

It is from this time, about 1,800 million years ago, that the atmosphere begins to look a little familiar. Before then, with no oxygen in the atmosphere, there was no layer of ozone (the form of oxygen in which three atoms, rather than the normal two, bind in each molecule) to screen out ultraviolet radiation, which would have reached the surface of the Earth in great quantities. This would have severely inhibited the spread of living organisms, for the radiation has a damaging effect on the forms of life which inhabit the Earth. But as oxygen spread, ozone would have been formed, by the action of the ultraviolet light on ordinary oxygen. Once sufficient ozone existed, the ultraviolet could no longer penetrate to the surface and with the surface waters and their rich sunlight now open to colonization, photosynthesis could speed up still further, soon establishing an atmosphere rich in oxygen.

Then things happened with startling swiftness. Presumably in response to the changed conditions, there was a burst of evolutionary activity which produced a diversity of multicellular animal life, within a space of about 100 million years. To gain some idea of just how quick this change was, in geological terms, it took about the same time as the formation of life-forms out of the chemicals of the early oceans, or the rapid diversification of the mammals during the Cainozoic (see chart on page 13), following the extinction of the dinosaurs. As for the dinosaurs themselves, they dominated life on Earth for something like 200 million years, starting in the final period of the Palaeozoic.

So the presence of life on Earth has had a dramatic effect on the evolution of the whole Earth system – or at least its outer layers. Today, we have a balance between the living and non-living parts of the whole system which has resulted from feedback between the two parts over thousands of millions of years. But, of course, the origin of life depended entirely on the physical conditions prevailing early in the development of the Earth.

Before life could begin, there must have been something around to act as food. Organic compounds are found in meteorites today and it seems likely that similar compounds were available from the earliest days on Earth. These compounds are more what we would regard as fuel than food; but even today there are organisms which live on this food – such as the kerosene used in jet aircraft (as operators of the aircraft have found to their cost). It is even possible to make use of these compounds in feeding people: we can now grow yeasts on oil, feed the yeasts to pigs and feed the pigs to people. This reversion of hydrocarbons to a vital role in the food chain does nothing, of course, to help the growing world energy crisis. If we can use oil reserves to feed our ever growing population, it seems particularly wasteful to use them as fuel for power stations and private transport. An understanding of the processes which mould our changing Earth may, however, alleviate some of the problems this raises, by locating untapped reserves of fuel (as we shall see in the next chapter).

So the first life-forms grew up to make use of the kind of food present early in the development of the Earth. It is hardly surprising that the remote descendants of those life-forms can still take in and use the same kinds of foods, at least at second hand. Obviously, with similar raw materials and suitable starting conditions life will form anywhere. 'Suitable conditions', in this case, include a supply of carbon and the energy to get the whole thing moving. Looking around the Solar System, the best places (apart from the Earth) where these conditions are met are not, curiously, on our near neighbours Venus and Mars, but on the giant planets Jupiter and Saturn. On these planets, according to the best theories, there are warm strata below the frozen top layers of atmosphere. Electric discharges – thunderstorms on a literally jovian scale – provide the energy required for organic compounds to be synthesized and there is plenty of carbon, in the form of methane gas.

How far evolution might have got under these conditions is still (just) in the province of science-fiction – but we may learn the answer from the next series of spacecraft to visit Jupiter and Saturn. However, it seems clear that man is pretty insignificant on the cosmic scale of things and that other forms of life as advanced as we are must exist on many planets in our own Galaxy, if not actually on other planets within our Solar System. It also seems that man is far from significant in the history of his own small planet, the Earth.

The history of man is so short compared with the history of the Earth that we can only understand our place on this planet by first considering a very broad timescale and then focussing on more recent events. The broadest scale of time divisions for the history of the Earth is provided by the divisions of classical geology; unfortunately, the names these eras bear have little relation to the periods they describe and sometimes they are very confusing. But they are now hallowed by traditional use, and no one seems likely to come up with an alternative system.

The period before 570 million years ago is called the Pre-Cambrian; since we know that the age of the Earth is about five *thousand* million years, 90 per cent of the history of our planet is lumped together under this unimpressive name. The reason, of course, is that the further back in time we look the less we know about what was happening to the Earth, and we know so little about the whole Pre-Cambrian that it would be silly to divide the period still further.

At least the name Pre-Cambrian is logical; as it suggests, the next period, from 570 to 500 million years Before the Present (B.P.), is called the Cambrian. Why Cambrian? Because rocks from this period are found in Wales, and Cambria was the Roman name for Wales. This is the kind of 'logic' which you need to unravel the mysteries of the geological names! But at least we know something about conditions on the Earth at that time. There was no life on the land, but all the main groups of invertebrates had already evolved in the sea – jellyfish, worms and sponges in particular. Seaweeds already existed in the Cambrian, providing food for these creatures. On the land – a barren, lifeless desert – few mountains had yet formed, but there was considerable volcanic activity.

From 500 to 435 million years ago the Ordovician period (named after a Celtic tribe) saw little change on land, but witnessed the emergence of vertebrates in the sea. However the next era, the Silurian (named after another Celtic tribe), saw a significant development on land. During the 40 million years of this period, mountain ranges were formed and, even more significant in the history of man, the first plants adapted to life on land. These were strange plants by modern standards, having no leaves as far as can be judged from fossil remains found in Australia; but they mark the first step in the long struggle of plants and animals to emerge from the comfort of the seas and conquer the land.

TIME BEFORE PRESENT (millions of years)	GEOLOGICAL ERA	GEOLOGICAL PERIOD	EPOCH
		Quaternary (3 million)	**Recent** (11,000) **Pleistocene** (3 million)
3	**Cainozoic** (About 1.5 per cent of Earth history)	**Tertiary** (67 million)	**Pliocene** (4 million) **Miocene** (18 million) **Oligocene** (15 million) **Eocene** (20 million) **Palaeocene** (10 million)
70	**Mesozoic** (About 3 per cent of Earth history)	**Cretaceous** (71 million) **Jurassic** (57 million) **Triassic** (32 million)	
230	**Palaeozoic** (About 8 per cent of Earth history)	**Permian** (55 million) **Carboniferous** (65 million) **Devonian** (50 million) **Silurian** (40 million) **Ordovician** (65 million) **Cambrian** (70 million)	
About 570	**Pre-Cambrian** (About 90 per cent of Earth history)		

Dates in brackets give the duration of each named time span. The dates given are approximate, and the further back they are in time the less accurately they are known. You may find slightly different dates quoted in other books but the broad sweep of events does not depend on pinning the dates of periods down to better than a few millions of years.

Following the plants, the first animals to leave the sea emerged during the Devonian (named after the English county), which covered the period from 395 million years B.P. to 345 million years B.P. They would have found the land a more turbulent place than it had been for millions of years, with extensive mountain building and volcanic activity – but perhaps these primitive invertebrates (millipedes, spiders, wingless insects and the like) were hardly equipped to notice what a dangerous place they were trying to invade!

In the Carboniferous (the Coal Age), which lasted from the end of the Devonian to about 280 million years B.P., reptiles appeared and became the first animals to breed on land. This is the period which begins to correspond well with the stereotyped view of prehistoric monsters lurking in swampy jungles, and this pattern continues through the Permian (named after a Russian province) and the Triassic (named after a mountain system in Germany), taking us up to 193 million years B.P. The Jurassic, which covered the next 57 million years, saw the dramatic development of flying animals, the dinosaurs which were the direct ancestors of modern birds. However in the next period, the Cretaceous (chalk), something even more dramatic happened; the non-flying dinosaurs disappeared after dominating the land for well over 200 million years. By 65 million years B.P., the end of the Cretaceous, placental mammals (whose young are nourished by the mother's blood until birth) had appeared; these were our direct ancestors. It is worth pausing here to reflect that the entire time span from the end of the dinosaurs to the emergence of our modern society is only one quarter of the time that reptiles were the dominant life on Earth.

As we move closer to the present, the geological time periods become shorter, reflecting the obvious truth that we know more about relatively recent events than about the distant past. At the end of the Cretaceous, it is time to refine our historical measures still further. From 65 million years ago to three million years B.P. is the Tertiary period, and the past three million years are lumped together as the Quaternary. However, to see man's place in the development of life on Earth we must switch from looking at these periods as a whole and examine the geological epochs within them. Remember, though, that the Tertiary and Quaternary combined (given the overall name Cainozoic) make up only one quarter of the time covered by the story of the reptiles, and only about one per cent of the entire history of the Earth.

The two epochs knowns as the Eocene and Palaeocene are usually combined in one geological unit, covering the time from 65 million years B.P. to 38 million years B.P. In that time, flowering plants became dominant and deciduous trees first rose to prominence; two groups of mammals (whales and sea cows) had the sense to leave the land and began to re-adapt themselves to life in the sea, while on the land ancestors of the elephant and horse, crocodiles, tortoises and, most significantly for us, monkeys and gibbons all put in an appearance.

The Oligocene takes us up to 26 million years B.P. and saw the beginning of the formation of the Alps. At last the geologist finds traces of man's immediate ancestors, the first primitive apes. In the next epoch, the Miocene (26 to seven million years B.P.), these apes began to spread, as did many other mammals including elephants. In the Pliocene, from seven to two million years ago, the great variety of mammal species began to thin out, but the man-like apes continued to develop and thrive to an exceptional extent. Ice ages killed off many species of plants and animals on land, leaving the hardier plants (oaks, hawthorn and willows among the trees) and the more adaptable animals (notably primitive man) to survive. By now, the sea contained life very much like that of today.

Thus the story of man began no more than 26 million years ago, and strictly speaking only two million years ago. So we must now start to look on an even finer timescale; taking millions of years at a time is too great a sweep of history.

Between two million and 500,000 years B.P. the kind of 'ape man' of the well-known picture emerged – Australopithecus, Java man and others. Over the next 450,000 years, species of these 'ape men' developed into early forms of modern man and of Neanderthal man, and not until a few tens of thousands of years ago did Cro-Magnon man begin to dominate – the beginning of the line *Homo Sapiens.* Ten thousand years ago modern man emerged. The history of agriculture and animal farming and the development of urban ways of life are all crammed into the past 10,000 years. The real history of man in his present form covers no more than 0.005 per cent of the time that reptiles dominated the Earth – and that period of more than 200 million years was no more than four per cent of the entire history of our planet.

So man really is a passing stranger as far as the Earth is concerned. Yet astonishingly, even with such a short history, man is in some ways on the edge of making a long-term mark on our planet, through interfering with the present balance of things. Pollution might, it has been argued, change the properties of the atmosphere drastically; nuclear warfare could be equally drastic and quicker acting. In either case, the results would be dramatic compared with events of the past 10,000 years and might lead to the end not just of man but of many other species. But even if you accept these predictions of doom, when you look at the consequences over millions and tens of millions of years they pale into insignificance. The Earth will go on, with or without man's interference, and it has already seen many changes far more dramatic than anything man might achieve, by accident or by design. What does a mere nuclear war matter to a planet that has seen whole continents created and destroyed? And what would be the significance of the end of man, compared with the part played by the reptiles in the history of the Earth? But a recognition of man's true insignificance is no cause to be disheartened. Just because our efforts are so puny compared with the forces of nature, it's very unlikely that man's activities will cause any terrible and irreversible changes in the environment. Changes, yes. But changes we can certainly learn to live with, just as our ancestors (unlike the whales) learned to live with the hazards of life on land rather than in the sea. And the size and long history of the Earth still provide us with a very large reservoir of natural resources to draw on. We hear a lot today about shortages of such necessities of modern civilization as oil and metal ore deposits; what that really means is that there are shortages of easily accessible and cheaply extracted reserves. Of course, the kind of profligate growth in consumption of such resources that we have seen in the twentieth century cannot continue for much longer. But the god of growth is clearly a false idol, and there is no need for more growth once everyone can be housed and fed reasonably. Even at present levels of consumption, known reserves of the resources now regarded as essential will last for a comfortable time, although it might be harder to extract them than we are used to (as in the case of developing oil shale resources rather than liquid reserves). But this is far from being the end of the story. The developing modern understanding of the history of the changing Earth can tell us two things: first, where there may still be reserves of untapped resources which will extend the time that society can continue in much the form we know it, and secondly, just how much (or how little) the *total* of such available reserves might be, so that we can, if the political will exists, budget those resources sensibly and plan ahead to ensure their best use. The Earth's resources certainly are finite; but geophysics and the new study of plate tectonics can tell us better than ever before how to make good use of the resources we have.

Chapter Two
New Sources of Fuel And Mineral Resources

Any new idea which can help us to locate and ration the remaining reserves of oil and gas around the world is of the greatest importance today. If the idea can also tell us something about other kinds of reserves, especially metals, it becomes doubly welcome. Today, modern theories of plate tectonics and continental drift provide just that kind of clue to making the best use of our dwindling resources. This kind of practical use of the new understanding of the Earth and how it has changed has followed from the 'revolution in the Earth sciences' in the 1960s, through which the concept of continental drift is now an accepted and fairly well-known feature of our understanding of the Earth. At the beginning of that decade, anyone who argued that the continents moved about on the surface of the Earth was regarded more or less as a crank; by the end of the 1960s, anyone who argued that the continents do *not* move was an object of ridicule in most geophysical circles. We shall see why this revolution came about later, but you do not need to know the background to the story to see the immense practical importance of the applications of the now proven fact of continental drift to problems of finding mineral reserves.

The most pressing problem in this line today is that of finding new reserves of fuel – oil and gas. This is also, in some ways, the most tricky problem to tackle, because these liquid and gaseous hydrocarbons can migrate, moving through porous rocks, and they also change chemically over millions of years, so that a reserve which once existed may be long gone by the time man comes on the scene. Even so, an understanding of continental drift can play a big part in the husbandry of these resources.

The hydrocarbon deposits are believed to have formed from organic material; oil comes chiefly from the remains of animal proteins, and gas (together with some oil) is produced by the decay of vegetation. Deposits are laid down in delta regions, where rivers deposit the organic debris in sediments that accumulate slowly over long stretches of geological time. Eventually, these sediments are covered by layers of rock and are shifted bodily by the processes involved in continental drift. The action of bacteria on the organic remains continues long after this, and produces reservoirs of oil and gas; this may evaporate, or flow away, or be changed still further by bacterial action. But in general the best chance of finding new reserves of hydrocarbons is to locate regions of the Earth's crust which were once the deltas of great rivers – and, equally important, geophysics can tell us confidently that regions which have never been part of such delta systems are not worth exploring, since they cannot contain any oil or gas.

It is easy to locate present-day deltas – but they are much too young for this long, slow process to have had time to convert their organic residues into fuel reserves. Old deltas are harder to find, and may even lie in present-day mountain ranges. The dispersed oil fields of the Andes, in South America, provide a good example; these probably formed when Africa and South America were still joined, before the South Atlantic Ocean opened and the two continents drifted apart. At that time, the Amazon river drained from east to west, emptying accumulated water from the one land mass of Africa and South America at what is now the west coast of South America. When the two continents separated, South America moved westwards, and this drift has played a large part in pushing up the Andes at the leading edge of the continent; the effect is rather like the folds of a tablecloth in front of a glass or any other object which is pushed across it. The result of this activity, in the case of South America, is that today the Amazon drains away from the Andes to the east, into the South Atlantic. In a few million years, the location of the present Amazon Basin will be a rich source of oil and gas, but today the right place to look for these reserves is in the mountainous region to the west, where the river used to drain.

The continental changes which provide the best insights have all occurred within the past 200 million years, since the time when all the land surface of the globe was concentrated in two giant continents, which geophysicists call Gondwanaland and Laurasia. For the 95 per cent of Earth history before the breakup of Gondwanaland and Laurasia the picture is less clear, because the geological record is confused by the upheavals caused when a previous generation of continents collided and fused together to form the two supercontinents. But that is of no great significance for these studies of oil and gas deposit locations, since the most productive present-day deposits were laid down only a few tens of millions of years ago. At that time, before the most recent ice age, the sea level was roughly constant for a long time, allowing many shallow deltas to become established and to deposit rich reserves of organic material. The shallow coastal seas would also have encouraged the formation of evaporites – sedimentary layers impervious to oil and gas, which have sealed off the reserves and kept them intact right up to the present day.

More than 90 per cent of known oil and gas fields are associated with evaporite deposits, which as their name suggests are produced when shallow pools or lagoons are heated by the Sun. Surface waters evaporate and dense brines form, which sink to the bottom of the pool where eventually the less soluble compounds in the water are deposited. The same kind of process occurs when shallow, flat regions bordering the sea are temporarily flooded, with evaporites being deposited as the floods retreat. It only needs occasional flooding, by the standards of human timescales, to produce impressive deposits over a period of a few million years. The conditions for production of evaporites are best in low latitudes where the Sun's heat is strongest – and the same latitudes are perfect for the optimum production of hydrocarbons through the richness of both animal and plant life. Briny pools also tend to be too saline for the taste of large creatures such as fish, which might eat the organic debris, while they are ideal for the micro-organisms (bacteria) essential for this debris to be broken down into hydrocarbons. All in all, the picture of how the deposits come to exist in the first place is nicely rounded out, with all the pieces meshing together. But the simple picture soon becomes complicated by the vagaries of geophysical processes. In particular, the rate at which hydrocarbons are produced from the organic raw material can be greatly speeded up by slight increases in temperature – and, on the other hand, a large rise in temperature could evaporate the oil completely, so that no reserves would remain.

Areas which have undergone such changes in temperature can be identified from an understanding of the processes of plate tectonics. When continents collide head on, mountains are created and great heat is generated in the rocks by the collision. The Himalayas were formed in this way, as the subcontinent of India ploughed into Eurasia, and in a collision as dramatic as this the heat generated deep in the rocks is more than enough to drive off any reserves of hydrocarbons. So there is no point in drilling for oil in the Himalayas, which is perhaps just as well. But sometimes continents collide more gently, or at a glancing angle, and these situations can produce the right amount of heat to speed up the development of oil and gas reserves without boiling them off altogether. This kind of situation is occurring today in northern Italy. Africa is moving slowly towards Europe, and in geological terms Italy is, in fact, an outstretched finger of Africa, and not part of Europe at all. The Alps (and further west the Pyrenees) are being thrust upwards by the squeezing effect as the Mediterranean closes up, and Italy is also twisting up against the underbelly of Europe. The result is that development of hydrocarbons from organic material has been encouraged by heating in parts of northern Italy, and geophysicists were able to predict, in 1973, that exploration for oil in those regions would be worthwhile. Almost two years later, moderate oil fields were indeed found in just the expected regions, providing the first direct confirmation of the value of this new understanding of how oil and gas reserves form. Just the same logic explains why there are gas fields on the margins of the Pyrenees – but these were already known in 1973, and it is always more satisfying to have a prediction confirmed than simply to explain a phenomenon which has already been discovered.

Another kind of heating, however, explains the location of many of the largest reserves of hydrocarbons. As we shall see later, the breakup of continents and the widening of oceans as they drift apart involves the creation of new sea floor in some places and its destruction in others. Both these extremes of tectonic activity are related to the heating of the rocks near by. Where new sea floor is being pushed outwards from the spreading centres of ocean ridges plenty of heat is generated – but, of course, the middle of the ocean is the last place to expect there to be any organic deposits which might be converted, with the aid of this warming, into hydrocarbons. The whole point about continental drift, however, is that the continents move; and as a result, from time to time, a continent overruns part of one of these spreading ridges. When this happens, something has to give, and either the spreading ridge will stop spreading or the continent will be split apart as the spreading continues. The detailed results are intriguing and well worth a careful study in due course, but all that matters here is that continental regions which have rocks containing organic remains slowly being converted by bacteria into hydrocarbons can be shifted bodily over the warming influence of spreading zones.

The heat flowing from these regions will cause the oil deposits forming to flow upwards into shallow rocks, and two of the greatest reserves yet exploited lie over just such zones. In Siberia, there are many highly productive gas fields sitting on top of a defunct arm of the spreading feature known as the Arctic Ridge; and around Los Angeles, in California, many productive oil fields are located close to the region where the so-called East Pacific Rise is splitting the Gulf of California, causing the earthquake activity of the notorious San Andreas Fault. Once again, the ideas of plate tectonics explain the presence of known reserves – but where, according to these modern theories, might there be as yet unexploited deposits of oil and gas?

The first thing we can learn is that these reserves may be more limited than has been guessed before. Because of the importance of low latitude locations for the formation of these deposits, much of the land now in the southern hemisphere is unlikely to contain a great deal of oil and gas. This is simply because the southern supercontinent of Gondwanaland spent less time at low latitudes than its northern counterpart Laurasia, so only small deposits are likely to have been laid down. In round terms, plate tectonics tells us that about two-thirds of the world's exploitable reserves of oil and gas lie in the northern hemisphere, which has already been extensively explored by conventional methods. The best sites for seeking untapped reserves in the southern hemisphere are in the south-west and western Pacific, and off north-western Australia. In the North, we have to turn to inhospitable regions to find sites where untapped oil and gas reserves may still lie, in particular the Labrador Sea, the Newfoundland Shelf, northern Greenland and the Canadian Arctic islands. No doubt the world's thirst for fuel will ensure that these rugged regions are exploited, just as the North Sea and Alaska have been. But this exploitation must take into account the fact that these are our last reserves of this kind, so that good husbandry will ensure that the best use is made of them.

With this cautious proviso in mind, North Sea oil can show us the sort of features which the oil explorers are now looking for. The picture in the mind of the explorer is of a shallow ocean with a flat coastal region, well watered by rivers carrying organic remains from the continental interior and preferably not too far from the equator, so that the Sun's heat can do an efficient job of drying out evaporites. The North Sea fits this image very well; it is a tantalizing situation for any geophysical Sherlock Holmes, since the oil deposits under the sea were in fact found after the discovery of gas fields in Holland encouraged offshore exploration, just before the geophysicists had enough information to predict the existence of the oil. Since the late 1960s, so much of the effort of oil companies has been concentrated in the North Sea (and Alaska, where it has been known for forty years that oil deposits exist) that no one as yet has made practical use of the powerful new tool of geophysical oil exploration provided by the new understanding of plate tectonics. But the giant companies have all carried out their own analyses of the new geophysics, and they all have their own ideas about where the next big oil discovery might be made. We don't have the resources of a giant oil company, but even a broad knowledge of plate tectonics gives us some intriguing clues.

Look again at the North Sea. This is the site of an old rift in the Earth's crust, which just failed to develop into a full size ocean. At the time when Europe and North America broke apart, such rifting took place over a fairly wide area before one rift began to develop fully and eventually became the Atlantic Ocean; indeed, in geophysical terms it was only by a hair's breadth that Britain failed to become an offshore island of Nova Scotia. So the obvious place to find other offshore oil reserves is on the sites of other rifts which failed to develop fully into ocean basins. Another clue comes simply from comparing the geology of opposite sides of the ocean, where continents were once joined. Salt basins along the Canadian coast, for example, hint strongly that early in the development of the Atlantic conditions at those latitudes were right for the development of oil deposits; so it seems logical to look for oil at the same latitude on the other side of the ocean, off the coast of Norway.

The location of rifts that failed to drift into full oceans is very promising. They are to be found up and down both sides of the Atlantic; near Europe we have, as well as the North Sea, the sea between Ireland and England, which might have been the spreading centre of the Atlantic if things had gone a little differently, and the English Channel also fits many of the characteristics of shallow seas suitable for oil deposits to be laid down. Further north there is an equally intriguing possibility, as a glance at any globe will show you. To the north and east of the Scandinavian peninsula lies the Barents Shelf, a large area of shallow sea with a similar tectonic history (over the past 100 million years at least) to that of the North Sea. Indeed, the whole region from northern Norway north and east past Spitzbergen is a flooded portion of the Eurasian continent, the biggest continental shelf in the world. This region of submerged continental crust is certainly bigger than Africa, and even the East Siberian Sea is essentially all flooded continent. Both these regions are inhospitable, but satisfy the requirement of shallow, low-lying regions well watered from the continental interior, which in an earlier phase of Earth history may well have produced evaporites leading to the presence of oil which remains trapped today. Maps centred on the North Pole show this very clearly – but these maps also show just what problems must be faced in getting the oil out, assuming it is there. The East Siberian Sea is an almost hopeless prospect for drilling wells today, since the limit of the pack ice stretches south almost to 70° N; in the Barents Sea, drift ice pushes well south of 75° N, and the climate generally would make North Sea drilling seem like a holiday in comparison. But closer in to North Cape something might be achieved once we become so desperate for oil that we are prepared to try drilling at sea north of the Arctic Circle. There is, of course, another problem, which will loom larger as the world becomes more desperate for new supplies of oil – the political one. The Barents Shelf is a sensitive region politically, and this may hinder development of its full potential. But in strictly non-political terms, it looks the best bet as the site most likely to follow Alaska and the North Sea into the big league of major new oil fields.

A region with no conflict of political interests and similar good prospects in geophysical terms is the Sea of Okhotsk, north of Japan and the Kuril Islands, and bounded on the east by Kamchatka. This looks almost like an upside-down North Sea (with Kamchatka the counterpart of Britain) and lies at much the same latitude but 150° to the east.

Searching through your atlas for sites of likely undiscovered oil reserves is a game that can provide hours of harmless fun; but remember, the big international oil companies are already playing that game in deadly earnest and have the geophysical expertise to narrow the choice down far further than is possible from a casual look at the present-day arrangement of continents and oceans. You cannot expect to form your own wildcat drilling company and beat the big boys to the next major strike. But, even aside from the long-term importance of oil for the future of our civilization, just a little knowledge could be of direct personal benefit to anyone who takes the trouble to understand the broad features of the changing Earth. If you should for some reason be determined to make an investment in the risky business of oil, look out for a company that is exploring or drilling in the region of one of these rifts that failed to drift, and remember the example of the North Sea.

What of other resources, such as metals? One of the most curious discoveries of recent years has been the existence of 'nodules' rich in nickel, cobalt and copper, which seem to litter the ocean floor in some parts of the world. Much has been made of the possibility of 'harvesting' these nodules, which are only a few centimetres across and usually lie at depths greater than 12,500 feet (3,810 metres), but the difficulties involved are immense, and the fact that the project is taken seriously at all shows just how desperate the search for these metals is becoming. The nodules are chiefly composed of iron and manganese (about twenty per cent on average), but they would not be worth harvesting just for these minerals since manganese deposits are still easy to reach on land, where they are also richer in manganese, and much the same is true of iron. However, the one or two per cent of nickel and 0.1 to 0.4 per cent of copper and cobalt in the nodules is enough to encourage serious efforts to get at them on a large scale. Even so, any 'mining' project (which would probably be more like a gigantic vacuum cleaning operation than any conventional mine) could only be a success if the miners are sure where the deposits can be found, and know in advance just how rich a particular 'strike' is likely to be. These are still formidable problems; it is far from clear how the nodules form, although there are significant differences between those in the Atlantic and those in the Pacific, and geophysicists are just about as baffled as anyone when it comes to predicting where on the sea bed they are most likely to be concentrated. It is early days yet to hope for any really successful exploitation of these nodules; but the commercial incentives which encourage investigation of this esoteric possible source of supply also make geophysical phenomena of great potential commercial importance rather better understood. As in the case of hydrocarbon reserves, if geophysics can tell us where mineral deposits are likely to be laid down then we can

85200

work out where those deposits are likely to have been carried to today. And, unlike oil reserves, deposits of minerals are unlikely to evaporate or run away once they have formed.

In describing how plate tectonics gives a guide to the location and extent of likely oil and gas reserves, it was helpful to talk in general, global terms. In the case of metal deposits, it seems more natural to concentrate on one specific example of a region where these deposits are being laid down today – in the Red Sea. This sea is an ocean in the making. As Africa moves away westwards from the bulk of the Eurasian continent to the east (and also swings northwards to close the Mediterranean) this zone of weakness is developing into a full-scale spreading region, with the associated heating that goes with this tectonic activity. The movement also seems to be producing the right conditions for ore deposits to be laid down at certain points along the bottom of the present Red Sea.

The first of these special regions – all of them are hot brine pools – was discovered only as recently as 1966. It has since become clear that in fact there is a whole chain of such pools running up the spine of the Red Sea. They contain water which is hotter than the surrounding sea, and also very rich in metallic salts compared with the bulk of the Red Sea waters – or indeed with the waters of the rest of the oceans of the world. The brines of the Red Sea pools have in them hundreds of times the concentration of dissolved materials usually found in seawater, and the sedimentary deposits underlying the pools are rich in heavy metals which include gold, silver, copper, lead and zinc. If these pools are found to be a feature of all enclosed basins associated with rift valleys, then it looks very much as if they are the sites where ore deposits are generally laid down. There is an important difference from the case of hydrocarbon reserves. These metal deposits do not need the action of bacteria over millions of years to become useful. The concentrations may be small, but they could be used without a great development of modern techniques if the need for the metals becomes great enough. Hot brine pools might well be a better proposition than sea-bed nodules for extracting rare metals.

However, the discovery of the first brine pool did not in

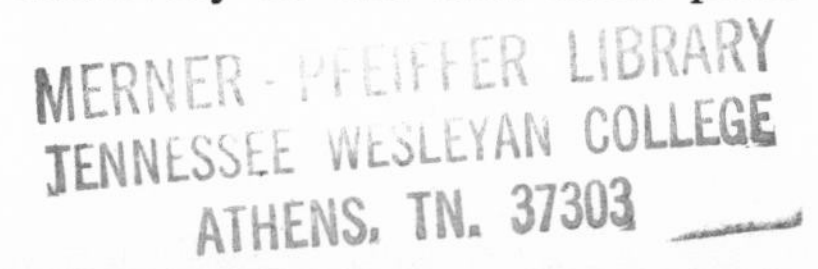
MERNER - PFEIFFER LIBRARY
TENNESSEE WESLEYAN COLLEGE
ATHENS, TN. 37303

itself open the door to exploitation. It was only in 1972 that a cruise by the Research Vessel *Valdivia* transformed the situation, when a German research team discovered thirteen brine pools, bringing the total then known in the Red Sea to seventeen. For the first time, it was then possible to make comparative investigations of the individual pools in order to find out which of their characteristics can indeed be regarded as typical of all systems like the Red Sea, and which might simply be peculiarities of the special local conditions around individual pools (Figure 2). From a geological point of view, it is now clear that there are at least two distinct parts of the Red Sea. The locations of the first seventeen brine pools discovered are shown in the map on page 29; the region between the Nereus Deep in the centre and the Suakin Deep to the south is one with a gently sloping sea floor across most of the width of the Red Sea. This falls away from a depth of about 500 or 600 metres into a steep-sided central trough about one kilometre deep, and the brine pools are very steep-sided depressions, more than two kilometres deep overall, lying inside this trough. Further south, however, the sea floor is made up of long, flat terraces, and this explains why no brine pools are found there – it is simply that there are no narrow and deep holes in which the brine can remain undisturbed by currents. And the most northerly deep yet found, Oceanographer Deep, may represent a third sub-division of the geology of the Red Sea.

In the steep-sided deeps the brines form layers of different salt concentrations, with the most concentrated (and therefore densest) layer at the bottom. The higher layers become gradually less concentrated until, at the top of the pools (but still one kilometre below the surface of the Red Sea) there is a transition zone which blends into the normal Red Sea waters. All this is easy enough to find, once you can locate a deep hole; in essence, the research ship just lowers self-sealing bottles to different depths, where they fill and seal automatically before they are hauled back to the ship for their contents to be analysed. But where do the metals in the pools come from?

There are two possibilities; one would imply that the pools are a feature of the tectonic activity associated with the

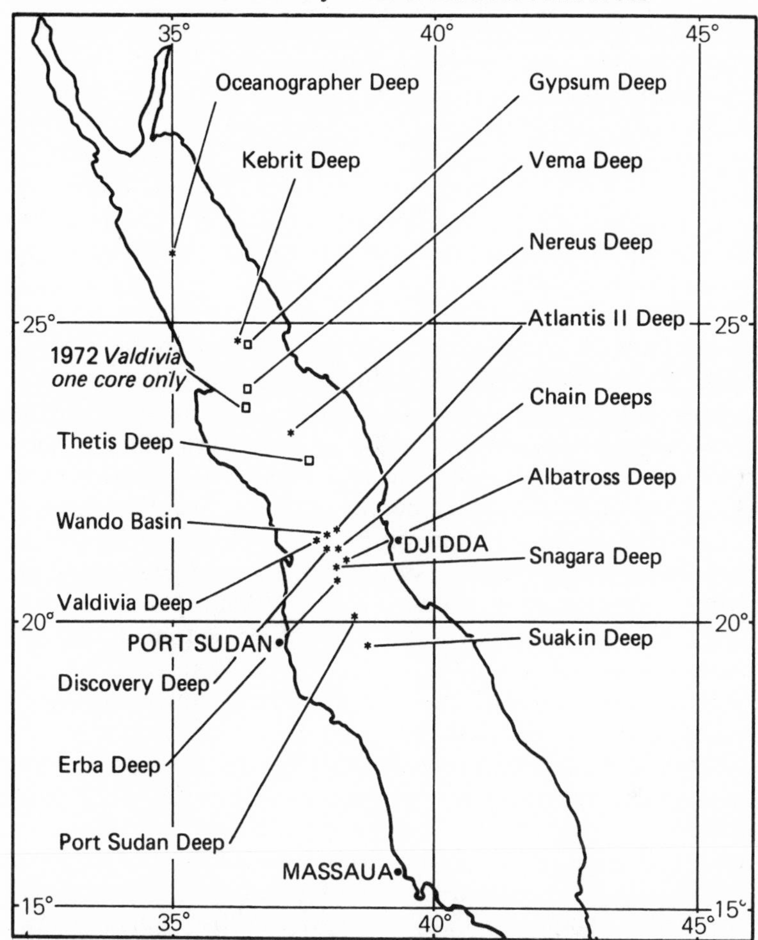

Fig.2 Distribution of brine pools and brine-derived sediments in the Red Sea. The map was prepared by H. Backer and M. Schoell in 1972; no more deeps had been discovered at the time of writing (October 1975). (Reproduced by permission of Dr Backer and *Nature*)

development of the embryonic ocean that is the present Red Sea, and the other that they are a more mundane feature arising just because the sea is shallow. Fortunately for our immediate descendants, it looks rather as if only Oceanographer Deep falls into the second category. The idea is that at some time in the fairly recent geological past the Red Sea dried out and has since been resubmerged. If that happened, then the water in the deeper holes of the trench would have evaporated slowly, leaving salt deposits which are now dissolving into the seawater which has re-entered the area. The more exciting possibility, which now seems to be established fact, is that the tectonic activity which is heating the pools, through gentle volcanic activity, is also introducing new metal salts into the pools. In this picture, the pools are not merely living off the deposits of their predecessors, but are laying down ever richer deposits, which might be taken advantage of by man.

Typical figures from an investigation of the Discovery Deep show the sort of conditions which exist in these pools. There, the salinity (which includes a measure of the salts of many different metals, not just the common sodium chloride) increases from the forty or so parts per thousand (four per cent) characteristic of seawater in the open ocean to more than 250 parts per thousand – and this increase takes place over an increase in depth of only a couple of hundred metres. At the same time, the temperature of the brine increases from about 22°C to about 44°C; this is a very strong clue that volcanic and other tectonic activity is warming the sea bed in this region from below.

The pools themselves seem to change their properties rapidly, indicating that the tectonic processes which are affecting them are vigorously active. The Atlantis II Deep, one of the first discovered, has been investigated by several research ships at different times since 1966. Between 1966 and 1972, the pool in the deep warmed from 55.9°C to 60.1°C, and the covering layer of water over the pool had warmed to about 50°C by the early 1970s. This indicates that a tremendous amount of heat has been put into the pool over a period of five or six years; according to calculations by a team from the Woods Hole Oceanographic Institute in Massachusetts, the increase in temperature of the pool, and the increase in the volume of water affected, must have been caused by an inflow of brine at a temperature of at least 113°C at the bottom of the submarine deep. Although this temperature is well above the boiling point of water at sea level, the pressure in the deep could keep it liquid even at a temperature of 360°C, so that poses no problem. The observations also show that the amount of iron and manganese in the brine of the Atlantis II Deep has increased since

1966; so the incoming hot brine must have a very high concentration of metals. Just where the inflowing water comes from is not yet clear, but it seems obvious that there is some kind of link between the pools and the volcanically active rocks beneath.

The Atlantis II Deep is a very special case. It is so hot that convection currents have been measured in its waters; as the water in the bottom layer is heated, it becomes less dense and rises up through the cooler layers. Because of this activity, the Atlantis II Deep is now spilling over into others near by; once the metal-rich brine reaches the top of the deep and cools, its high density carries it along the sea floor near by. But even pools which are not connected to one another by this kind of convective activity over the intervening sea floor have such similar concentrations of metals and nearly identical temperatures that it seems they must be connected, presumably by tunnels below the sea floor. That in turn seems to point the way towards an explanation of where the sudden influx of hot brine into the Atlantis II Deep has come from, since if these tunnels were heated by an influx of volcanic magma (molten rock) the water in them would become very hot, and would also absorb large amounts of metal salts from the magma. The theory looked good. But the only way that the research teams could be absolutely sure that these pools are not the result of simple evaporation processes was to find a pool with different properties, which fitted the evaporite picture better than pools such as Atlantis II Deep. This happened in 1972, when a team from Imperial College in London investigated Oceanographer Deep.

This deep lies slightly outside the central valley of the Red Sea, at 26° 16.6′ N, 30° 01′ E. Just to the east and west of the deep the sea floor is at a depth of 960 metres, while the bottom of the deep itself is at 1,528 metres; the width of the depression is only 0.6 nautical miles at the 1,350 metre depth contour, and 0.4 nautical miles at the 1,450 metre contour, with an overall length of less than two nautical miles (when scientists are put on board ship, the resulting measurements seem to come out in a curious mixture of scientific and nautical units!). Samples of brine from the deep, and sediments drilled from the sea bed, have been analysed to find out the origin of the Oceanographer Deep. The amount of metals such as iron and copper found in the brine is more than in normal seawater, but less than the concentrations found in active pools such as the Atlantis II Deep. In terms of weight, normal seawater contains 0.02 milligrams (mg) of iron in each kilogram (kg) of water; Atlantis II Deep contains 80 mg per kg and the Oceanographer Deep contains 0.2 mg per kg. According to the research team from Imperial College who made the measurements, the core drilled from the bottom sediments was rich in organic material, and gave off an extremely strong odour of hydrogen sulphide (the 'rotten eggs' smell). In more quantitative tests, they found that the distribution of metallic elements in the core was very similar to the distribution in cores of this kind drilled from Norwegian fjords.

All this evidence suggests very strongly that the Oceanographer Deep, the most northerly of the brine pools yet found in the Red Sea, is indeed an effect of the region having dried up and then been resubmerged. The deposits of salt that result – evaporites – are like some found recently in cores from the bed of the Mediterranean, which indicate that that sea has been repeatedly dried out and resubmerged; perhaps the drying out of the northern part of the Red Sea has been related to the changes in the level of the Mediterranean.

For the geophysicists, this discovery of a second kind of brine pool in the Red Sea was the crucial evidence needed to show that their understanding of the formation of the brine pools is correct. They have produced two theories to account for the formation of such pools, and now the tests in the Red Sea have shown that there are indeed two kinds of pool, each fitting nicely into the picture which results from one of the theories. It is just because the Oceanographer Deep is so clearly an evaporite pool, and because the deeps further south are so clearly different from Oceanographer Deep, that the active pools can be identified with such certainty with active tectonic processes – the processes of continental drift – going on in the region today. This understanding of how mineral deposits are laid down is at an earlier stage of development than the understanding of how oil and gas deposits form. As yet, no one has taken this work to the extent of making general predictions about the likely locations of ore deposits, unlike the comparable situation of using continental drift to locate fuel reserves. But if you want to become a prospector with hopes of striking it rich, the first thing to do is to consult a geophysicist with a good grasp of the new ideas of plate tectonics; he should be able to predict where in the world tectonic activity in the distant past may have produced ore deposits which we can mine if not today then certainly in the very near future. Geophysics and plate tectonics are studies of great practical importance to the problems facing the world today; this is no ivory tower research. In the rest of this book, I shall try to outline the broad picture of how this new science of the 1960s developed, and what it means to our understanding of planet Earth. The outline may not be sufficient to enable you to find your own private nickel mine, but it should remove some of the mystery from the arguments and counter-arguments used in the debate about man's significance as an inhabitant of one small planet.

Chapter Three
Old Ideas about the Earth

No concept of the idea of continental drift could come in anything like its present form, until man had discovered that the Earth is round and had located at least the general outlines of the major continents. Even then, it is one thing to notice in a general way that the bulge of the western coast of Africa looks as if it might once have fitted in to the gap between North and South America and quite another matter altogether to develop an understanding of how and why the two continents came to have this peculiar configuration. In 1620, Francis Bacon wrote of the New World and the Old: 'Each region has similar Isthmuses and similar Capes, which is no mere accidental occurrence. So too the New and Old Worlds are Conformable in this, that both Worlds are broad and extended towards the North, but narrow and pointed towards the South.' Did Bacon believe in continental drift? We shall never know; but this example shows that speculation about the causes of the present arrangement of the continents has been rife since the earliest days of reliable mapping on a global scale.

The earliest records we have of man's thoughts about his place in the Universe come from the oldest civilization, that of China. The fame of the ancient Chinese as astronomers, and especially as observers of supernovas and comets, is well merited; it is less well known, perhaps, that the Chinese philosopher-scientists developed theories of earthquakes, and instruments to record these shakings of the Earth, long before anything comparable was developed in the West. Of course, they had some incentive for this kind of study, since China is in a region where seismic disturbances are common.

But, even with this incentive, little progress towards what we would regard as realistic theories could be made until there was a reasonable understanding of the nature of the Earth. Some of the earliest ideas conceived earthquakes as the result of the balance of forces between the Yin and the Yang, with seismic troubles occurring when the Yang has been subordinated, or imprisoned, by the Yin. These concepts developed into what might be called a 'pneumatic' theory of earthquakes, developed quite independently in almost exactly the same form by the civilizations of the Mediterranean at about the same time, a couple of millennia ago.

Other Greek ideas suggested that earthquakes were caused by water bursting through into hollows inside the Earth or, conversely, by large masses of the Earth drying out after being saturated by water, with lumps falling down into caverns in the interior. The ideas all look equally silly in the light of modern knowledge, but we should not forget that these ancient thinkers did not have any modern information about the Earth and that their ideas were a serious attempt at understanding the Earth in the light of the best knowledge available at that time. What the inadequacies of these ideas really highlight is the need for an understanding of the Earth as a whole before we can comprehend how the overall workings of the Earth system can produce local effects such as earthquakes. The story of the development of ideas about the makeup and workings of the Earth is, then, closely related to the story of the exploration of the Earth. Leaving aside the lost knowledge of the Norse explorers (and the even more thoroughly lost knowledge of their possible predecessors on the sea paths to the New World), this story really began in the fifteenth century, when Europeans began to explore westwards and southwards by sea. Christopher Columbus, the most famous of these European seafarers, ironically owes his place in history to a thorough piece of self-deception regarding the true nature of the Earth.

Long before Columbus's time, about 230 B.C., Eratosthenes of Cyrene calculated the circumference of the Earth and arrived, as much by luck as judgment, at a sum close to the correct figure – about 40,000 kilometres. His calculations depended on measuring the length of a shadow cast by a vertical rod at Alexandria at noon on the day of the summer solstice when, he believed, the Sun stood exactly overhead at Syene on the upper Nile (present-day Aswan). The measurements showed that the circumference of the Earth must be fifty times the distance between the two cities, which Eratosthenes knew at least approximately. Modern astronomers can point out errors in each step of these calculations – but, by lucky coincidence, in the end they all cancelled out!

Later estimates of the Earth's size were not quite so lucky, and the estimate enshrined by Ptolemy in his writings, which were accepted almost as gospel by most scholars for a full 1,500 years, was close to 33,000 kilometres, only four-fifths of the true value. Ptolemy's *Geographica* was translated into Latin only at the beginning of the fifteenth century; with the addition of new information from the travels of Marco Polo, and by juggling the figures still further to fit in with his hopes, Columbus was able to convince himself – if not all scholars of the time – that the lands of the Far East lay within the range of ships sailing westwards from Europe.

According to Ptolemy, the distance from Portugal to the far side of Asia was about 16,500 kilometres, which is about 1½ times too far. Adding China and Japan on to the limits of Ptolemy's geography gave a further 4,780 kilometres, adding up to a total of two-thirds of the size of Ptolemy's globe. Columbus managed to fudge the figures still further by picking out the calculations of another astronomer, Alfragenus (a ninth-century Moslem), and using these to 'correct' Ptolemy's estimate of the circumference of the Earth down to 30,000 kilometres. Columbus was not the first – and certainly far from the last – person to find evidence supporting a treasured theory by picking and choosing from the available data. But at least he was prepared to risk his own life when the time came to test his theories.

When all the fudging of figures was done, Columbus reckoned to have shown that Japan lay only about 4,300 kilometres west of the Canary Islands – at about the longitude where we in fact find Cuba. Although even

Columbus would hardly have set out on a voyage of 19,000 kilometres, the real distance between Europe and the East by the westward route, it is difficult not to escape the conclusion that whatever theories had been around at the time he would have found a way of pulling the distance he wanted out of the hat. The rest of the story is, of course, familiar history. Ironically, to his dying day (in 1506) Columbus believed not only that he had opened up the westward sea route to Asia, but even that on his fourth and last voyage he had all but reached the mouth of the Ganges river. But others soon appreciated the truth – that a New World had been discovered. The first map showing the new lands separated from Asia by another large ocean, and giving the new lands (or at least the southern continent) the name America, was drawn in 1507 by an Alsatian, Martin Waldseemueller. It took a further 450 years, until the middle of the nineteenth century, for seafarers to find all the continents, including Antarctica. With the growth of geographical knowledge, there was a growth of geological and geophysical speculation, and this formed the springboard for the very recent development of the modern theory of plate tectonics.

The science of geology began, after a fashion, when sea shells were discovered on dry land, with the implication that the surface of the Earth has not always been as it is today. This discovery caused much confusion and theological argument at first. On the one hand, it was argued that the shells were proof of the reality of the biblical flood; on the other, it was seriously suggested that they might be artifacts 'planted' by the Devil with the object of confusing the development of man's inquisitive intellect! As further investigations of changes wrought in the surface of the Earth seemed to point to a very long history of development, there was increased, sometimes bitter, conflict with the Church authorities. Compared with what we now know about the history of our planet, summarized in Chapter Two, it is a sobering thought to remember that it was only 300 years ago, in the mid-seventeenth century, that Archbishop Ussher put forward his famous calculation of the date of the Creation, showing that Heaven and Earth began on the night before Sunday, October 23rd in the year 4004 B.C. One important consequence of this school of thought, endowing the Earth with a very short history, was that the only way in which dramatic changes in the face of the Earth could be explained was by invoking dramatic catastrophes, such as the flood. Until a true picture of the age of the Earth emerged, there was no way in which geological changes could be interpreted in terms of long, slow evolution of surface features into their present forms.

One seventeenth-century scientist who rejected the catastrophic interpretation of changes in the Earth was Robert Hooke. He saw the presence of fossils of sea creatures on land as evidence of cyclic changes from sea to land and back again over long periods of time, and suggested that new rocks might form from the effects of heat below the Earth's surface, or by other processes. Furthermore, he realized the link between vertical movements of the crust and seismic activity (in fact, he suggested that the British Isles rose up from the sea as Atlantis sank, in a sort of seesaw balancing act). Hooke also discussed polar wandering and climatic changes in a curiously modern way; but although his geological ideas appeared in print in 1705 (after his death), they did not gain any great acceptance at the time, and seem to have been either forgotten or ignored. It was another century before the rest of the geological world caught up with Hooke's ideas.

Even at the beginning of the nineteenth century the few geologists who argued that the Earth must be thousands of millions of years old went largely unheeded. The great explorer Alexander von Humboldt suggested in some detail how the Old and New Worlds had been separated by the coursing effects of the waters of the biblical flood, rushing from south to north and carving out the Atlantic Ocean as nothing more than a river valley on a grand scale. But von Humboldt did draw attention to the many similarities between the continents on the east and west sides of these oceans, although he never interpreted the similarities in terms of continental drift. The presentation of geological arguments highlighting the similarities of the continents on either side of the Atlantic had to wait a further half-century, until

1858, when Antonio Snider, an American working in Paris, published his work *La Création et ces mystères devoilés.* This work presented the first published fitting together of the Atlantic continents prior to their separation (see Figure 3), which was used to explain the similarities between fossils found in coal deposits of Europe and North America. Snider still subscribed to the theory of a separation occurring in a catastrophic event associated with the flood; and his fit of the continents differs in some respects from modern reconstructions of the situation before the breakup of the one supercontinent which once existed on the Earth. But, from now on, although proponents of the idea of continental drift were to go through a long period of struggle before the idea gained acceptance, the possibility of continents having separated from some earlier supercontinent was at least discussed. Snider's map marks the beginning, in this sense, of geological recognition of the existence of the concept, if not of its acceptance. That acceptance was to take another hundred years.

One version of the catastrophic 'drift' of the continents was particularly appealing. In a sense, this idea – that the Moon was thrown off from the Earth early in its history and that this process tore apart the early supercontinent – is still very appealing; but today, thanks partly to studies of rocks brought back from the Moon, we know that the Earth and Moon formed separately and were never part of one planet in the way envisaged by some nineteenth-century

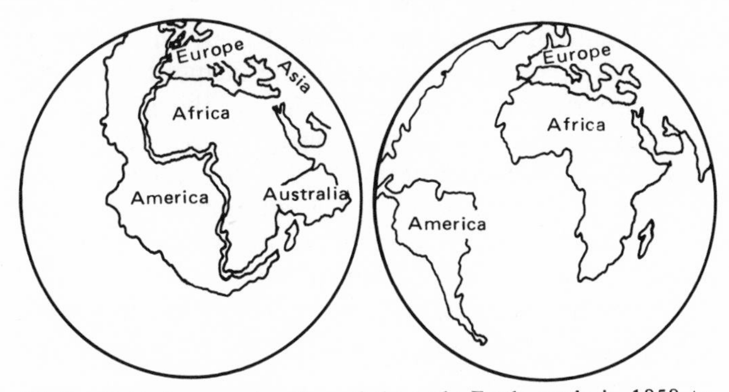

Fig.3 Snider's reconstruction of the early Earth, made in 1858 to explain similarities between fossils found in Europe and America.

scientists. In its most complete early form, this idea was described by Osmond Fisher in *Nature*, on January 12th, 1882. He built upon the earlier speculations of George Darwin (a son of Charles), who calculated that if the Earth is slowing down steadily through the influence of solar tides, then about 57 million years ago a situation might have occurred in which the rotation period (then about five hours) would be in resonance with the solar tide, producing conditions just right for the primitive Earth to split into two unequal parts. Fisher suggested that a great rent would be torn out of the Earth's crust on one side, centred on the equator. This rent – the Pacific Basin – would be partly filled as molten rock from the interior welled up and solidified, but as Fisher put it:

> There would, however, have necessarily been a certain amount of flow in the upper fluid layers towards the cavity, and this would have carried the cooled granitic crust which, floating on it, still remained upon the earth along with it. What was left of the granitic crust would therefore be broken up into fragmentary areas, now represented by the continents. This would make the Atlantic a great rent, and explain the rude parallelism which exists between the contours of America and the Old World.

Appealing though the idea is, we now know that the Moon was a separate body at least 4,000 million years ago and has evolved essentially as an independent body in its own right. But Fisher's speculations contain one element that provides an early pointer to the development of the idea of continental drift – the idea of the continents as granitic masses floating on a molten interior and being carried along by currents in these fluid layers. The alternative view – widely held by the geological establishment at the turn of the century – was that the continents did not move laterally, but that ocean basins had grown up through the sinking of continents (or 'land bridges') which used to link the present continents.

Once again, the idea has a certain superficial appeal. In its most thoroughly worked-out form, the model envisaged a cooling, contracting Earth. As the interior cooled and shrank, so it was argued, great blocks of the crust could fall inwards to fill the gaps, creating ocean basins (the theory provides a curious nineteenth-century echo of some of the ancient Chinese speculations about earthquakes!). By the late nineteenth century, there was a pressing need to account for the similarity of fossils found in the Old World and the New, and land bridges provided an obvious possibility. In addition, of course, the mystique of such legends as that of lost Atlantis had as much appeal to our Victorian ancestors as it has today. The overall concept of Earth science developed from the idea of a cooling, contracting Earth was presented in a very complete form by the pioneering geologist Eduard Suess; but all of these ideas were based on one assumption which soon proved incorrect. The Earth is not simply cooling down from an initial hot state – which would imply that it formed only a few tens of millions of years ago! – but is still receiving heat from an internal source. This source is the decay of radioactive elements in the Earth, and it is radioactive decay which has helped to keep the Earth's interior warm enough for it still to be molten after not tens of millions, but thousands of millions, of years. Once again, we see that an understanding of the changing face of the Earth depends not upon some specialist field of study, but on a synthesis of important developments across the whole spectrum of the physical sciences – and, if you include studies of fossils and of variations in the kinds of animals in different continents today, then the spectrum must be extended to include biological sciences as well.

In the first quarter of the twentieth century, two American geologists, Frank Taylor and Howard Baker, independently produced their theories of continental drift, and over the same period the German Alfred Wegener, who is today regarded as the 'father' of the concept, put forward his own ideas on the subject. The reason for Wegener's place in history is simple: unlike other proponents of the various versions of the continental drift hypothesis, Wegener kicked up a fuss about his ideas, rather than just publishing them and leaving them to be accepted or not by the geological community. In addition, as an astronomer turned meteorologist turned geophysicist, Wegener had a broader grasp of the Earth as a whole than most people, including most modern geophysicists; and perhaps, therefore, he had less to lose by sticking his neck out in support of the concept of continental drift.

Seemingly, Wegener became interested in the possibility of continental drift as early as 1910, five years after he received his doctorate. His thoughts appeared in a small monograph in 1915, which was revised in 1920. But it was the third edition of this book, published in 1922 and translated in 1924, that formed the basis for the often furious discussion about continental drift in the second quarter of this century. It could hardly be said that all of this discussion was rational and scientific. Many opponents of Wegener made no attempt to justify their opposition, but simply stated that his theories were unbelievable; others found fault with details of his ideas, and made the wild leap of concluding on this basis that the broad sweep of the Wegenerian concept of continental drift was wrong. Outside the main geological argument, some biologists and geographers were willing to accept the evidence Wegener put forward showing the similarities of ancient plant and animal life in what are now separate continents, but were swayed by the emotive arguments against the drift hypothesis into reviving the idea of land bridges to account for the similarities. Yet, amidst all the confusion, Wegener's summary of his hypothesis (presented in the 1922 edition of his book, and recently quoted by Ursula Marvin in her excellent historical review *Continental Drift**) contains many features now incorporated in the modern theory of plate tectonics. According to Wegener:

1. The continents and ocean floors are fundamentally different. The continents are seen as blocks of light granitic rock (sial, from silica-alumina) floating on denser basaltic rock (sima, from Si-Mg), which is exposed as the ocean floors.
2. The sial has grown smaller in extent and its thickness has

* Smithsonian Institution, Washington, 1973.

increased over geological time by folding, while it has also split into fragments, which now cover one-third of the Earth's surface.

3. The present-day continental blocks still have essentially the same outlines as they have had since the breakup of a supercontinent in the Mesozoic.
4. Bringing in his wide-ranging knowledge of different aspects of Earth sciences, Wegener confirmed the reconstruction of this supercontinent by matching mountain ranges, sedimentary formations, evidence from the scars of ancient glaciations and the distribution of both fossil and living plants and animals. It is this broad sweep of evidence which makes it so remarkable that Wegener's version of continental drift failed to gain acceptance in the 1920s.
5. Wegener went further than his predecessors by pointing to rift valleys as signs of incipient continental breakup, indicating that the process is continuing today. He explained mountain ranges as, in many cases, the result of crumpling of the leading edges of the continental blocks as they move through denser material like icebergs through the oceans.
6. Wegener supported the view, from climatic evidence, that as well as the drift of the continents there has been a movement of the poles (polar wandering), by which the Earth's surface has moved through 4,000 kilometres relative to the poles since the Permian.
7. Wegener did not invoke a shrinking or expanding Earth to provide the driving force for continental drift, but explained the process in terms of rotational and tidal forces acting slowly but surely over geological time.
8. Direct measurements of latitude and longitude showed, said Wegener, that certain land areas, most notably Greenland, were moving at measurable rates.

Where Wegener really fell down, perhaps, was in his proposed mechanism for continental drift. The external forces just cannot account for the separation of continents, producing movements in many different directions for millions of years on the surface of a spinning planet. Harold Jeffreys, of the University of Cambridge, made this problem abundantly clear in the 1920s, and dealt with it in his classic book *The*

Earth,* which is still in print. (Jeffreys never revised his views, even when continental drift became, as plate tectonics, the fashionable geophysical study of the 1960s.) In 1928, at a symposium held by the American Association of Petroleum Geologists, Wegener also came under attack because he was an outsider, not one of the geological community, whose ideas must therefore be intrinsically suspect. The death of Wegener a few years after this symposium left the theory of continental drift without a champion, and little more was heard of it for decades. Ironically, at the very time that the controversy seemed to have been resolved, if only by default, in favour of the traditionalists, Arthur Holmes of the University of Edinburgh proposed a mechanism for the drift which overcame Jeffreys's objections. This involved an internal process, convection currents driven by radioactive heat, which is similar in many ways to the ideas accepted today. But it was not acceptable in the 1930s, and although in retrospect all the evidence necessary to convince geologists of the reality of drift was available at that time, it was another thirty years or more before the establishment view was changed.

Why did it take so long? Were the discoveries of the 1960s – which we shall learn more of later – really so dramatically convincing? I think not. Quite simply, the geologists of the 1930s seem to have been unable to accept the overturning of their applecart, which would have meant discarding a wealth of carefully accumulated concepts. The story goes that one participant at the 1928 conference actually said, 'If we are to believe Wegener's hypothesis, we must forget everything that has been learned in the past seventy years and start all over again,' as if that in itself was a good reason for rejecting the hypothesis! Within a few years, it was being seriously argued that Wegener's idea was so ridiculous that it should not even be mentioned to students as an hypothesis, because it might cause them confusion – yet, today, it would be difficult for any geophysicist to obtain a teaching post in a university if he did not profess a belief in the concept of continental drift!

* Cambridge University Press.

It is worth remembering, too, that the models of the convective processes which are now believed to drive the movement of the continents are still far from being completely satisfactory today, so that argument alone cannot account for the reversal of opinion. It seems, basically, that the Earth scientists of the 1960s were simply more ready than their counterparts of the 1920s and 1930s to discard old dogma. The temptation to link this changed attitude towards traditions with social changes over the same decades is strong, but outside the scope of this book. But there were developments in techniques and in geological knowledge which at the very least helped the idea of continental drift to become established in the more receptive environment of the 1960s. Before we look at the details of this theory, in its modern incarnation as plate tectonics, it makes sense to take a look at just how geophysicists and other Earth scientists obtain their information and build up an overall picture of the world as it is today.

Chapter Four
Developing the New World Picture

Our understanding of the Earth as a planet has grown up through many areas of specialist study – geology, meteorology, oceanography, astronomy and others. But the coherent picture depends on putting the different areas of knowledge together: the non-specialist approach. Some of the pieces of the picture are more important than others, and some seem less important at present simply because we do not yet have the knowledge to make their true significance intelligible. In the case of the oceans, for example, it is easy to see that the seas play a vital part in the development of the Earth, and through their interactions with the atmosphere must profoundly influence the climate. But it is far from simple to understand the oceans, and as yet oceanographers have only scratched the surface of a vast area of research.

The atmosphere is perhaps a little better understood than the oceans, and its importance to the whole Earth system is more obvious. There would not be any Earth scientists to speculate about the changing face of the globe if we did not have a suitable atmosphere for them to breathe! Both atmosphere and oceans play an important part in changing our planet, particularly through the process of erosion. When we look at an ordinary river in an ordinary valley – let alone the Grand Canyon – it is a sobering reflection on the speed with which man's works pass away to realize that the whole valley has been carved out of the surface of the Earth by the action, perhaps over thousands of years, of a fairly insignificant stream of water. The wind, carrying abrasive particles of dust, also erodes the Earth's surface, wearing down whole mountain ranges over geological time. And as we have seen, it is the rivers of the past that have laid down the organic deposits so valuable to us today as sources of energy. It would certainly be wrong to dismiss lightly the importance of wind and water to the changing Earth. But, even so, all this is something in the way of being the icing on the cake. The atmosphere and oceans owe their existence to the presence of the solid Earth – and they probably were produced by the outgassing of material from the solid rocks, as we shall see later. The main scope of geology and geophysics, which has produced the new understanding of continental drift in terms of global tectonics, is essentially concerned with the solid (or, in its interior, semi-solid) Earth. But even here there are many separate parts which must be related to produce a coherent whole.

Geological techniques are concerned with first obtaining samples of rocks and the minerals they contain, and then examining the samples in the laboratory and interpreting their significance. The specimen collection process ranges from the time-honoured technique of chipping bits of rock off an outcrop with a hammer to obtaining core samples from the ocean bed with a boring tube. One of the most important aids to the geologist is the presence of the fossilized remains of dead plants and animals in different layers of rock.

When a sea creature dies, it sinks to the bottom of the ocean where its remains form part of the ooze from which, in the fullness of geological time, solid rocks form and are lifted to the surface, perhaps to be incorporated in new mountain ranges. Similarly, leaves or ferns may fall upon swampy ground on the land and be preserved as fossils. Once a species of life dies out, through the natural processes of evolution, it is never found again in a younger rock; and since no two species are identical, no new species is ever the same as an extinct one. So the relative age of any rock layer (or of many strata) can be determined by comparing the fossils found by geologists. With the development of an understanding of radioactive processes, geologists have a new method for dating rock strata, by measuring the decay of radioactive elements found in traces in the rocks. A combination of these techniques has produced the geological timescales of the chart on page 13 and has given us a good grasp of the true age of the Earth.

If geology is primarily concerned with the status quo of the Earth's crust, geophysics can be thought of as the study of the changing Earth and of the Earth's interior. The two aspects of geophysics are inseparably interlocked. The most important technique of the geophysicist is still seismology, the study of the Earth's interior by means of low-frequency sound waves produced by explosions or by earthquakes. It is easy enough to record the sound waves, which produce a gentle shaking of the rocks that can be turned into a trace on a chart or graph of seismic activity. But interpreting these seismic waves is a problem of great complexity, which Sir Edward Bullard, one of the pioneers of the concept of plate tectonics, likened to the problem of trying to deduce the structure of a grand piano by listening to the noise it makes when pushed down a flight of stairs. But today the seismological technique can be spectacularly successful, as in a study of the island of Cyprus made in the early 1970s by British geophysicists. The survey showed that the Troodos Massif of Cyprus is probably an upthrust fragment of oceanic crust, squeezed up from the sea floor, like toothpaste squeezed out of a tube, by the forces of plate tectonics. This study is of more than just theoretical importance, since it explains the origin of the great copper deposits which gave Cyprus its name and have made the island economically important from the time of ancient Greece right up to the present day. So the story is certainly worth more than a passing look.

The idea that the mountains which form the Troodos Massif might have been part of the sea bed in the not so distant geological past has in fact been around since the 1950s. A massif is a block of the Earth's crust which is markedly different from surrounding rocks, and generally such blocks are very rugged in topography. But during recent geological history, a massif has developed as a single unit. In the case of the Troodos, there is evidence from magnetic surveys and from conventional geological studies of the rocks that the mountains have an oceanic origin. But the final proof came from the seismic survey, in a crucial experiment which depended on determining the pattern of rock strata below the Troodos mountains by recording the echoes of shock waves from explosions fired at the bottom of water-filled boreholes.

In different layers of rock, the sound waves from such explosions travel at different speeds, and the boundaries between rock layers provide particularly good echoes – it's much the same as the way in which submarines can be detected by echo sounders (sonar) on board ships. In order to map the layers of the Earth's crust, it is necessary to have a chain of listening stations at known distances from the sites of the explosions. When the recordings from each station are analysed and all the information put together, it is possible to build up an accurate picture of the underlying rock formations – provided the stations have been sited in the optimum locations. Unfortunately, the rugged terrain of the Cyprus mountains forced the seismic surveyors of a few years ago to make do with second best in some of their choices of recording locations, but even so the picture they have built up strongly suggests that the rock strata under the Troodos Massif are like those of the normal ocean floor.

Many of these seismic refraction experiments have been carried out at sea, and they show that underneath a layer of sediment, usually about two kilometres thick, are two rock layers before the underlying mantle is reached. In the lower of these layers, the seismic velocity of shock waves is about 6.7 kilometres per second, and in the upper layer it is close to 5.1 kilometres per second. The Cyprus survey showed that the massif has a three-layer structure, but the surface layer is a relatively recent addition, made up of volcanic lava, and is only poorly defined. So it cannot be regarded as a truly separate layer, and should be discounted in the same way that the layer of sediment is discounted in studies of the structure of the rocks of the sea floor. In the first 'true' rock layer of the Troodos, the seismic velocity turned out to be between 5.1 and 5.3 kilometres per second (the uncertainty is caused by the difficulty of locating the recording stations in ideal sites); the second layer produced particularly clear reflections, indicating a velocity of 6.4 kilometres per second. These tally with the figures for the rocks underlying the oceans – and the accord is even better when allowance is made for the great weight of water pressing down on the sea floor, which will squeeze the underlying rocks and thereby increase the velocity of sound waves moving through them to a slight extent.

This kind of study is of great value, because in the Troodos geophysicists are actually able to touch and examine at close range what they now know is really a piece of oceanic crust. Exploitation of the resources of the sea floor is one of the few ways still open to mankind to tap new wealth, and it is going to be a lot easier to tap that wealth – and to reassure the people who must provide the economic backing – now that geophysicists can have the benefit of detailed studies of the Troodos, and any other uplifted fragments of the sea floor that they can discover. But it is still essential to develop ways of observing the sea floor directly, and two techniques in particular demonstrate the adaptability of modern technology in meeting the needs of the geophysicists and geologists who are the pathfinders for what will eventually be the economic exploitation of the ocean bottoms.

One of these techniques involves taking the seismographs out to the sea floor, instead of restricting seismic observations to the dry part of the globe. The particular importance of this is that although recordings of man-made explosions can tell us a great deal about the upper layers of the Earth – as in the case of the Troodos – information about the deeper layers of the Earth's interior must come from studies of natural processes, such as earthquakes. As the study of our planet has developed, it has become possible to obtain useful information not only from large earthquakes, which produce a reaction on seismographic equipment around the world, at permanent recording stations, but also from much smaller events, or micro-earthquakes. This is doubly valuable, since there are many more such events than the artificial explosions used for special localized surveys; micro-earthquakes on land can now be studied by efficient, cheap and portable seismograph stations, which can be set up near particularly interesting regions of the Earth, such as the line of the notorious San Andreas Fault in California. The frequent micro-earthquakes thus provide information about local conditions without the bother of drilling holes and setting off explosions, and this local colour is complementary to the broad picture of the Earth as a whole provided by studies of large seismic disturbances.

However, many of the most interesting features of the Earth, in terms of the ideas of plate tectonics, are hidden beneath the oceans. In particular, the ocean ridges which run through most oceans are sites of great activity where new oceanic crust seems to be being created, pushing out on either side of the ridge as the ocean basin widens. This kind of activity can be studied conveniently in Iceland, where part of the Atlantic Ridge actually pushes above the surface of the sea. For the great majority of the ocean ridge system, however, micro-earthquake activity associated with the spreading of the sea floor can only be studied by depositing seismometers *in situ* on the sea floor.

Various groups around the world have tackled this problem. In the U.S.A., developing new geophysical tools is a valuable industry and firms such as Teledyne Geotech, based just outside Washington D.C., keep the details of their tools very much a trade secret. In other countries, such as the U.S.S.R., some of the professional cards are played equally close to the chest, for reasons of national economy. But there is no secret about the three basic approaches to the development of submersible seismographs. First, a seismograph sealed in a watertight container can be lowered to the sea bed on a cable and information transmitted over the cable to a ship above. This technique works well in shallow, calm waters, and has been used by Soviet seismologists investigating the Black Sea and the Baltic regions; but it has the great disadvantage that any movement of the cable connecting the ship and instruments will cause a vibration which could overwhelm the vibrations produced by micro-earthquakes. This makes the technique useless in the open ocean.

An alternative approach is to drop a seismograph to the sea bed complete with an underwater transmitter so that information is relayed to the ship above without any physical link. But this requires complex and expensive electronic equipment, which is used once and then left to rot on the sea floor. And, as with the cable technique, it ties an expensive research vessel to one spot while the recordings are being made. So the most promising approach to providing seismic surveys of micro-earthquakes on ocean ridges is to drop a completely self-contained seismograph and recorder which monitors seismic activity for a set time and then floats to the surface where it is recovered. The recorded information can then be analysed at leisure.

Put like that, this last technique sounds simple and obvious. But each step involves painstaking work. First, the information can only be of real value if the exact position of the monitoring station on the sea bed is known. So this kind of unit contains an acoustic 'pinger' which operates during the fall to the sea bed and for a short time afterwards. By tracking the pings and taking navigational fixes from artificial Earth satellites, the crew of the parent ship can locate the seismograph on the sea bed to an accuracy of 0.1 nautical miles. The pinger would, of course, affect the sensitive seismograph, so it is turned off before the survey proper begins. As well as needing to know the exact location of the recorder, the seismologists need an accurate measurement of the time at which disturbances begin, so a submersible seismograph station must carry its own crystal clock and make timing 'marks' on the magnetic tape used for recording the seismograph traces. Finally, after a pre-set time the buoyant seismograph station is released from its heavy base and floats to the surface for recovery, which is aided by a direction-finding radio beacon which switches on at the surface.

Although rugged and reliable, these devices are not extravagantly expensive; each costs several thousand dollars. One research ship can drop a chain of them along a region of interest, steaming back along the chain later to pick up the recording stations, remove their tapes, and use them again. It is through direct tests of this kind that the resources of the sea floor, as well as its geophysical secrets, will be revealed in detail.

Investigation must also be made into the kind of rocks

that are on the sea floor, especially when it comes to deciding the best sites for prospective oil wells and the like. Today, it is no longer necessary to go through the laborious process of drilling a test core for analysis and repeating the drilling many times across a region of interest, at least when all that is required is the broad outline of the underlying rock structures. The principle behind the quick, new method of mapping the sea bed is the measurement of the radioactivity of the rocks under the sea. The elements potassium, uranium and thorium are all slightly radioactive and the proportions of these elements in different rocks provide a characteristic 'fingerprint' by which the rock can be identified. Granite, for example, is particularly rich in uranium, although not radioactive enough to cause any harm to the inhabitants of towns perched on granite hills.

Once again, it seems simple to study the radioactivity, perhaps by trawling an ordinary geiger counter over the sea floor. But as in the case of undersea seismometers, the practical details complicate the issue. A gamma ray detector in a steel cylinder would certainly do the job of measuring radioactivity, but if it were simply towed behind a ship it would catch on every snag, wreck and rock outcrop; as trawlermen know from bitter experience, this almost inevitably results in the tow being broken and the package on the end being lost. The answer this time is to enclose the whole towing cable in a rubber sleeve, like a thick hosepipe, which has the same diameter as the steel cylinder on the end. Then, there are no sharp corners or edges to catch on projections on the sea bed, and the whole thing (which has been dubbed the 'electric eel') can be towed across the sea floor at three or four knots with no risk of damage.

During such a survey, radioactive particles which enter the steel cylinder cause sodium iodide particles in the gamma ray detector to emit pulses of light. These are converted into electrical impulses by a detector inside the 'eel' and travel along an electrical cable inside the towing cable directly to the ship above. There they are fed into a computer which counts the number of pulses of each magnitude; the magnitude of the pulses is characteristic of the radioactive particles striking the 'eel' below, and thus provides an indication of the kind of rock encountered.

Several traverses, a few miles apart, are needed to build up an outline geological map of a region; the quick survey can then be supplemented by conventional drilling techniques to provide a complete picture. In the waters of the continental shelf – of greatest importance to the economic future of so many countries – the 'electric eel' has made it possible to determine the composition of submarine rocks quickly and easily.

Larger scale (remote) surveying methods provide geophysicists with an insight into two features of the Earth which are of key importance not for localized purposes (such as deciding where to drill an oilwell) but for the overall world picture. Because different rocks have different densities, the gravitational pull of the Earth varies from place to place, to an extent measured in millionths of the overall average pull of gravity on the Earth's surface. These small variations can be measured by instruments called gravimeters, which work on the same principle as a spring balance, with a weight attached to a very sensitive hair spring made of fine wire or quartz. A pointer indicates the deflection of the weight by local gravitational anomalies, and although they are very sensitive gravimeters can also be made very rugged, so that they can be flown in aircraft to provide an overview of the gravity characteristics of a region. Alternatively, portable gravimeters can be carried in back packs to points of specific interest, but this more laborious approach is rapidly being outmoded by development of better methods of airborne gravimetric surveying. This kind of survey provides information about regions several hundreds of miles across, such as the North Slope region of Alaska which is now known to harbour large quantities of oil.

A very similar approach is used to study magnetic anomalies, which are observed using magnetometers flown in, or trailed behind, aircraft. This time, it is a compass-type needle that is held in place by a fine spring and deflected by changes in the magnetism of the rocks below. For the theory of plate tectonics and sea-floor spreading, magnetic surveys of the oceanic crust have been of key importance, as we shall see in the next chapter. The detailed study of rock magnetism depends still on the traditional method of obtaining samples of the rock for examination in the laboratory, but aerial magnetometer surveys do have the great advantage of showing up very clearly rocks which are rich in metallic ores, such as the vast ore fields of the Australian outback. In addition, they can point the way for more traditional studies of particular regions of the oceans which seem in some way peculiar, as in the case of studies of the Rockall Bank in the late 1960s and early 1970s.

Rockall, it turns out, is rather like the case of Cyprus in reverse. Whereas the Troodos Massif is formed of ocean floor squeezed above the sea, the sea floor around Rockall is characteristic of the rocks normally found in continental masses. A magnetic survey of the Rockall Bank disclosed a belt of magnetic 'anomalies', typical of the kind which are associated with lava flows, which sweep around the island in a broad curve. The pattern suggests that an extinct volcanic centre underlies the Rockall Bank, and seismic reflections indicate that rather than the two-layer sea-floor structure characteristic of oceanic rock, the thicknesses of the rock layers around Rockall are typical of those under land masses. Following these remote surveys, geophysicists obtained samples of rock from the sea floor near Rockall, to test the theory that the island and bank may be a fragment of continental crust, left behind as the Atlantic Ocean opened.

Almost all rocks underlying oceans are relatively young, geologically speaking, since they have only been laid down during the period that the oceans have been opening up. But some samples from the Rockall Bank are 1,000 million years old, five times older than most of the rocks of the continental crust – let alone the sea floor. Comparably old rocks are found in very few places on the Earth, but notably in parts of Canada. The age is determined very reliably from argon dating, which depends on the fact that as radioactive potassium in rock decays argon is released and trapped in the crystalline fabric of the rocks; the amount of argon present today is an excellent measure of how long it is since the rock was molten, its geochemical age, and shows unambiguously that Rockall is not a typical sea-floor region.

It seems that when a continent starts to break up under the influence of whatever deep-seated driving forces power continental drift, the break may not be entirely clean. A small portion of the continental crust may be split off from the main mass and remain surrounded on both sides by the new rocks which form the ocean floor; in addition to the Rockall Bank, which now looks to be a leftover fragment of eastern Canada, such 'microcontinents' are found in Madagascar and the Seychelles Bank in the Indian Ocean. This kind of study provides impressive confirmatory evidence that continents do move as the ocean basins open up through the processes of plate tectonics, and it plays an essential part in the methods by which the new world picture has been developed. But the initial breakthrough in the 1960s, which alone changed the climate of geophysical opinion in favour of the concept of continental drift, came from magnetic studies of the ocean floor. The breakthrough came when these studies revealed that the sea floor is spreading outwards from the ocean ridges, pushing the continents apart, and providing evidence not just that the continents *can* move, but that they *must* move.

Chapter Five
Sea-floor Spreading, Continental Drift and Plate Tectonics

During the 1950s, seismic studies at sea showed that the Earth's crust is much thinner in the ocean basins than it is under the continents – the underlying mantle of the Earth's interior is only 5 to 7 kilometres below the sea, but on average the thickness of the continental crust is about 34 kilometres, and in places it is 80 or 90 kilometres thick. The other great discovery about the sea floor which was made at much the same time was that the Earth's surface beneath the waves is just as rugged as the land surface, with mountains, submarine canyons and in particular the ocean ridges which I have already mentioned. The ridge in the middle of the Atlantic Ocean stands some three kilometres above the plains of the sea floor, and this and other topographical features show the sea floor to be an active geological region. The ridge in the North Atlantic was the first to be studied in detail and is still the best documented; perhaps fortunately for the development of geophysics in the 1960s, it is also the ridge which fits most simply into the new world picture.

Along the centre of the ridge runs an active rift valley encompassing many sites of volcanism. In 1960, Professor Harry Hess, of Princeton University, explained this and all the other newly discovered features of the sea floor in terms of the first model of sea-floor spreading, reviving the idea of continental drift which had been rejected for so long. According to this model, the ocean ridges are produced by upwelling convection currents in the fluid material of the Earth's mantle. These currents carry mantle material to the surface at the ridge, then 'spread' it outwards on either side, pushing the continents apart and forming young ocean basins from the solidifying mantle material. This idea carries the Wegenerian concept of continental drift a significant step forward, by providing a driving force for the drift. But it raises several new questions.

The first puzzle is to explain what happens to all the new oceanic crust that is being created. Straightforward geological dating of the breakup of the continents which now lie on opposite sides of the Atlantic implies a spreading rate of about one centimetre a year on either side of the ridge – a widening of the Atlantic by two centimetres every year. Such a rate of creation of new crust would produce all the present-day ocean floor in only 200 million years, less than five per cent of the age of the Earth. There are only two ways to resolve this dilemma. Either the Earth has expanded by two-thirds in that short time, or oceanic crust is being destroyed at other sites around the world as rapidly as it is being created at ocean ridges.

It seems very unlikely that such a rapid expansion could have occurred without producing other, even more obvious, effects and in particular the shape of the present continents suggests that they have always existed on a globe the size of the present Earth. This can be seen very simply by making cut-outs of the main continents on a globe and sliding them about to see how they may once have fitted together. The fits are so good that it seems certain that even before the supercontinents broke up the Earth was indeed the same size as it is today.

This leaves the possibility of the destruction of oceanic crust, and Hess suggested that the deep trench systems at the edges of some oceans, particularly the western Pacific, mark the sites where the downward limb of a convection zone is sweeping material back into the mantle, where it melts and, perhaps, eventually goes through the whole cycle again. This is still the basic process envisaged in the modern version of the concept of sea-floor spreading. Circumstantial evidence also lends weight to the model. There are only relatively thin layers of sediment over ocean basins, showing that they have only existed for a small fraction of geological time; no very old rocks have been recovered from the oceans, except in such special cases as the Rockall Bank, where their presence is actually required by the best models of continental drift. Most impressive of all perhaps to the non-specialist, the greatest concentration of global seismic activity occurs along the ridge crests and trenches, just where, according to Hess, the sea-floor spreading process is producing most activity (see Figure 4). Even with all this supporting evidence, however, it proved difficult to convince the geophysics community of the reality of sea-floor spreading and continental drift. It was not until 1963 that the vital magnetic evidence – which in retrospect seems the clinching data – was presented in print, and even then, according to some geophysicists closely involved with the problem at the time, it was not immediately realized just what the new evidence implied, when linked with Hess's ideas of sea-floor spreading.

One of the two or three people who pioneered the modern concept of continental drift, in its incarnation as plate tectonics, is Dr Dan McKenzie of the University of Cambridge. McKenzie recalls that it was a talk Hess gave in Cambridge in 1962, when McKenzie was still an undergraduate, that first fired his imagination and set him puzzling over these problems. Others also seem to have been motivated by this talk, and a year later Fred Vine and D.H. Matthews, then rather more senior members of the Department of Geodesy and Geophysics at Cambridge than McKenzie, described the relation of magnetic anomalies – magnetic 'stripes' on the sea floor – to the ocean ridges. It is not clear who first realized the possibility of linking Hess's sea-floor spreading idea with the occurrence of these magnetic stripes, although McKenzie recalls Vine discussing the possibility with him around that time. Today, the link seems obvious; but we must remember the weight of opposition to continental drift which still existed in the early 1960s, and the perhaps natural reluctance of some geophysicists to become tarred with the same brush as Wegener. It may seem that the transformation of the Earth sciences in the 1960s was a revolution, but for the first few years of that decade at least progress was made only slowly, in spite of the rapid accumulation of new evidence.

The magnetic stripes were first observed in the north-east

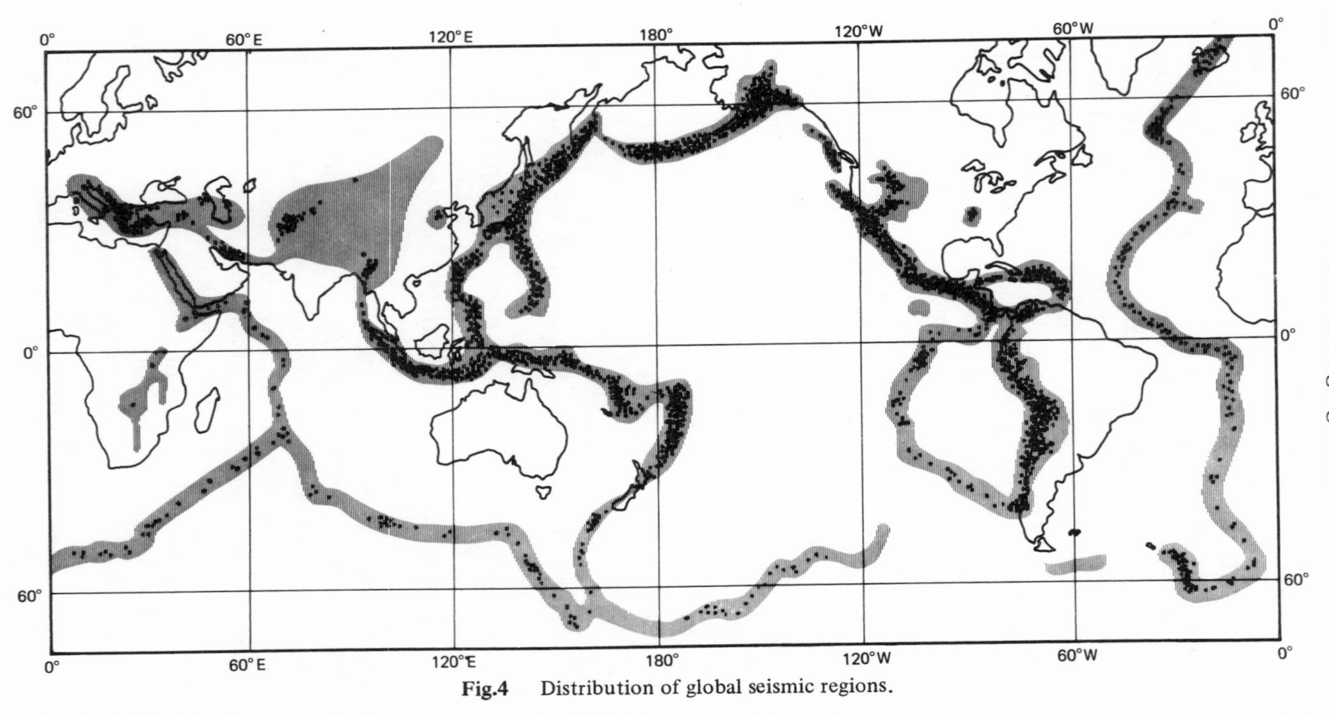

Fig.4 Distribution of global seismic regions.

Pacific, when magnetometers were towed behind ships to obtain magnetic surveys of the ocean bottom. The stripes run north–south, and show up because in one stripe the direction of magnetization of the underlying rock may be the same as that of the Earth's field today, while in the next stripe the magnetization is reversed, and so on. Now, if the ocean floor has formed by the solidification of molten rock from the mantle, it is natural that it should have some 'fossil magnetism', because the rock will be magnetized by the Earth's field as it sets, to produce a weak permanent magnet. But, like a bar magnet, once the rock has solidified its magnetism will be fixed. If some rocks today have the opposite magnetism to that corresponding to the Earth's present-day magnetic field, it must mean that either the rocks have turned over since they formed (without melting in the process) or that the Earth's magnetism has changed since they were laid down. The first possibility was so completely unlikely that the second alternative – repeated reversals of the Earth's magnetic field – had to be taken seriously. It also meant that the adjacent stripes of sea floor must have been laid down at different times, just as the sea-floor spreading hypothesis required.

It is worth emphasizing just what this discovery means in terms of the Earth's magnetism. In Sherlock Holmes' tradition, we have eliminated the impossible idea that solid rocks of the Earth's crust have been turned over bodily in adjacent stripes, and we are left with the improbable, but true, alternative that the whole magnetic field of the Earth repeatedly changes direction completely, with north and south magnetic poles interchanging. Further studies of fossil magnetism have now shown that during such a changeover the field first dies away to nothing, then builds up again in the opposite direction (so that compass needles, for example, would point to the opposite pole). Just to confuse the issue a little, sometimes it seems that the field dies away and then builds up again in the *same* direction as before. The whole process takes a few thousand years – the mere blink of an eye in terms of geological timescales – with periods between reversals lasting for anything from 100,000 to 50 million years. During the past 70 million years or so reversals seem to have been relatively frequent, occurring once or twice every million years, and since the Earth's field seems to have been weakening for the past 150 years at a rate which would reduce it to zero in 2,000 years time, we may even be experiencing such a reversal now.

So the overall picture of sea-floor spreading sees new crust being laid down at ocean ridges and magnetized by the Earth's field. As the crust spreads out on either side, the Earth's magnetism undergoes sudden reversals from time to time, so the magnetization 'fossilized' into the rocks also changes. The youngest rocks, next to the ridge, are magnetized in accordance with the present-day field; as we look further away from the ridge, we are essentially looking back in time, thanks to a kind of magnetic tape recorder, and seeing how the Earth's magnetism has varied. One obvious prediction of this model is that the pattern of magnetic anomalies must be symmetrical about the ocean ridges if newly formed crust has spread out evenly on either side – and that is exactly what we do find (see Figure 5). But in 1963 it was still a subject of debate whether or not the Earth's magnetic field had ever reversed, and, more important, even the supporters of the then new ideas of sea-floor spreading and the magnetic tape recorder could not produce a definitive chronology, dating the occurrence of such geomagnetic reversals throughout Earth history.

The situation slowly improved as more magnetic surveys were carried out and as the potassium-argon technique was developed to date lava flows accurately. It took until 1966 for the timescale of geomagnetic reversals to be developed reliably for the past 3½ million years, and this is the timescale that has been used in Figure 5 to indicate the chronology of sea-floor spreading processes. The timescale has since been improved and extended further into the past by studies of rocks on land, where successive layers of rock are magnetized by the Earth's magnetic field as they solidify. The vertical pattern showing the same geomagnetic reversals as those needed to explain the magnetic stripes of the ocean floor nicely rounds off the picture of the Earth's ability to switch magnetic poles on occasion (and very rapidly by the standards of geological timescales). How and why these

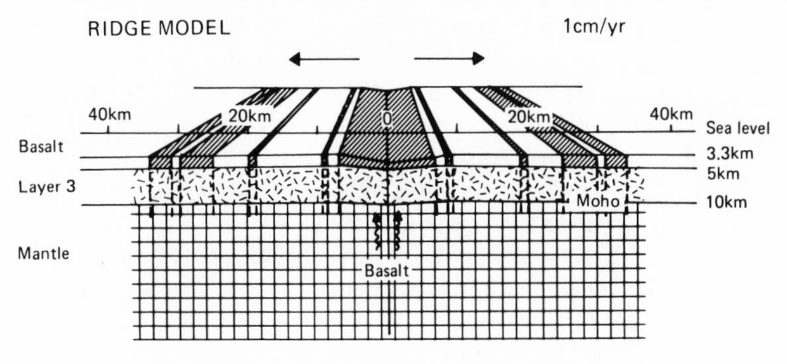

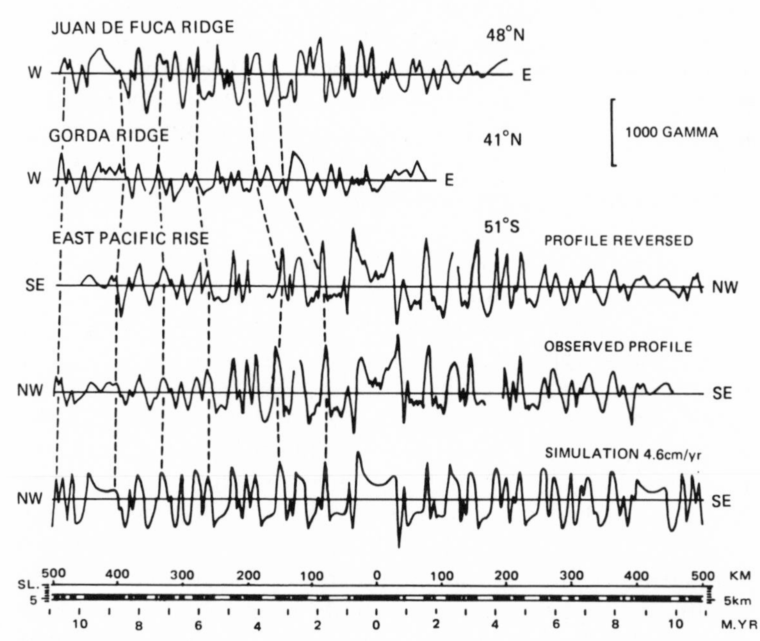

Fig.5 Spreading-ridge model, with dark and light stripes indicating opposite magnetization. (From *Understanding the Earth*, reproduced by permission of Artemis Press Ltd)

geomagnetic reversals occur is still a problem for geophysicists to resolve, but the fact that they do take place is now seen as unarguable proof of sea-floor spreading, for how, other than by a combination of this spreading and the geomagnetic reversals, could the magnetic stripes of the ocean floor be formed?

Quite remarkably, however (at least in retrospect), it was not this overwhelming new evidence which swung the bulk of geophysical opinion behind the idea of continental drift in the mid-1960s, but a restatement of the old ideas, in terms almost indistinguishable from those of Wegener. The exact occasion when opinion was first seen to change is remembered by many geophysicists: at a symposium on continental drift sponsored by the Royal Society in 1964, Sir Edward Bullard, then the head of the Department of Geodesy and Geophysics at Cambridge. presented a map showing the fit of the continents on either side of the Atlantic (Figure 6). This reconstruction was the first widely publicized fit achieved with the aid of a modern electronic computer, and it was accepted where earlier, almost identical, hand-drawn maps had not been. This is a curious puzzle; but first let us see just how these reconstructions are made.

As you can see from the contour lines in any atlas, the exact coastlines of the continents would shift noticeably if the sea level were a little higher or lower than it is today. Such sea-level changes have occurred frequently through alteration in the amount of water locked up in the polar ice caps, and perhaps through other geophysical processes, so it is not really appropriate to produce reconstructions of past supercontinents by fitting together the outlines of present continents marked by present-day coastlines. Fortunately, however, there is a much better guide to the true edge of a continent than the present coastline, and that is provided by the edge of the continental shelf. At this edge, the sea floor plunges away steeply from shallow depths down to the four kilometres or so typical of the ocean floor, and it seems sensible to argue that if continents have broken apart then this edge marks the break. Taking the 'coastline' half-way or a quarter-way down this steep slope (at the two kilometre or one kilometre contour), we obtain shapes for the continents – pieces of a global jigsaw puzzle – which do indeed fit together well. This was clear even in 1958, when a fit between South America and Africa very similar to that of Figure 6 was produced. That fit, however, was produced by moving the pieces of 'jigsaw puzzle' about on a globe to obtain the least overlap by eye, and although the human eye

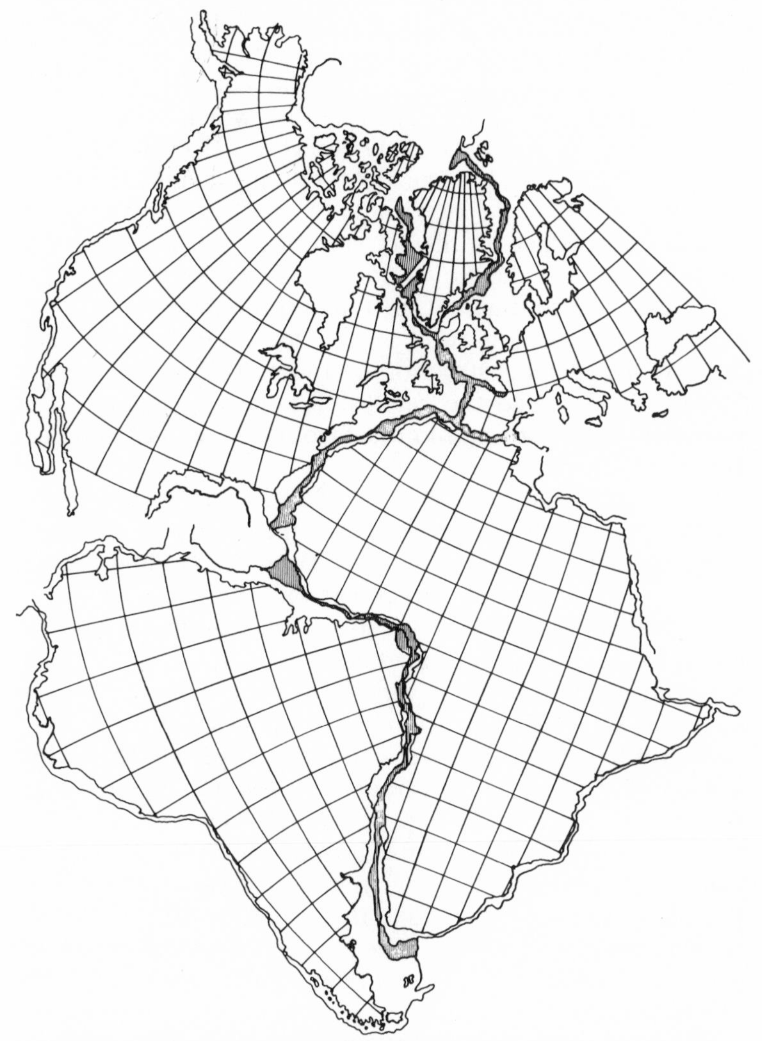

Fig.6 Bullard's fit of the continents prior to the opening of the Atlantic.

is really a rather good judge of shapes, it was only when the work was duplicated with the aid of an electronic computer that it seems to have been widely accepted.

In 1964 the electronic computer was still very much a new toy for most scientists, and it is tempting to speculate that it was a primitive respect for the power of the new toy that encouraged the geophysical community to accept the reconstruction presented by Bullard and his colleagues where it had rejected previous reconstructions which today seem equally convincing. It is likely, however, that Bullard was simply lucky in striking the right tide in the affairs of geophysics. In 1964, the new ideas of sea-floor spreading and the magnetic stripes on the ocean floor were being discussed; the very fact that the Royal Society should be sponsoring a symposium on continental drift shows the dramatic shift of establishment opinion from even a few years before. The computer-aided reconstruction of the past fit of the continents simply provided the catalyst for the expression of the rising groundswell of opinion in favour of the once heretical idea of continental drift.

Throughout the rest of the 1960s, a great weight of evidence was turned up to confirm the reality of drift and spreading. The advent of the U.S. Deep-Sea Drilling Project JOIDES (from Joint Oceanographic Institutions Deep Earth Sampling) was one of the most significant developments. A chain of drilling sites across the Mid-Atlantic Ridge produced samples of rock which could be dated accurately by the new techniques of modern geophysics; these showed that spreading is symmetrical about the ridge, and that it has been occurring steadily throughout the Cainozoic at an almost uniform rate of two centimetres a year, exactly in line with the estimates derived from the magnetic stripes. Further studies in this programme, and various pieces of geological evidence, now make it possible to date the break-up of various pieces of the former supercontinent of Gondwanaland and the beginning of the opening of the present oceans.

Today, we can see an embryonic ocean opening in the Red Sea, as already mentioned (page 27); the northern end of the North Atlantic and the Arctic Ocean have opened up only within the past 100 million years (during the Upper Cretaceous and Tertiary), following the opening of the South Atlantic, which began in the Lower Cretaceous, more than 100 million years ago. But all these studies emphasize that both the terms applied to the process – 'continental drift' and 'sea-floor spreading' – are inappropriate since the changing face of the Earth affects both oceans and continents. It was the incorporation of both processes into one unified model, the theory of plate tectonics, that marked the end of the revolution in the Earth sciences and the beginning of a new established order. The beginning of the revolution had come with the realization that the Earth's crust can be deformed drastically at ocean ridges and trenches – the boundaries of what are now known as the 'plates' of the Earth's crust, which give the new theory its name. The completion of the revolution came with the emphasis that whatever happens at the plate boundaries, there is little or no deformation within the plates, which behave as solid islands jostling each other in a changing global jigsaw puzzle pattern.

The concept is so familiar today that it is difficult to appreciate just how novel plate tectonics was just a decade or so ago. Indeed, it was only in 1967 that the term was first used in a geophysical paper, when Dan McKenzie and his colleague Dr R.L. Parker published what has become a classic article in the scientific journal *Nature*. In this, they explained features of the North Pacific in terms of the movements of rigid plates sliding over the surface of a sphere. Both the authors were then working in California, although McKenzie was soon to return to Cambridge, where he was one of the prime movers in developing the concept so rapidly that, in his own words, 'by the end of 1968 the theory was essentially complete', at least in its broad outlines. Thus the prime interest of geophysicists today is in applications of the theory of plate tectonics, in an attempt to explain the detailed workings of our planet and to solve more immediately practical problems, such as tracking down new resources.

The classic 1967 paper by McKenzie and Parker brought together all the new ideas of geophysics in one coherent theory. They showed how the geophysical observations could be explained if the seismically quiet areas (see Figure 4) move as rigid plates, and these plates interact with each other at boundaries defined by seismically active regions. In 1967, they applied the idea in detail to the North Pacific area, showing that it worked for more than a quarter of the

Earth's surface, and in 1968 Dr Jason Morgan, of Princeton University (who had been working along the same lines independently), produced a world map showing the seismically quiet areas as six main plates (essentially the quiet regions shown in Figure 4) and twelve sub-plates. Then, the theory of plate tectonics was essentially complete. In spite of superficial similarities, it is radically different from the old idea of continental drift: we no longer think of the movement of continents through the oceanic crust like icebergs through the water, but of the interactions of a few, large plates on the backs of which continents can ride.

The most interesting regions of geological activity today, therefore, are around the margins where two or more plates meet. There are three types of margin. Constructive margins are found where new oceanic crust created at ocean ridges spreads away on either side, so that we see two plates moving away from one another; destructive margins, on the other hand, occur at the deep trenches where one plate slides underneath the edge of another, diving at about 45° back down into the mantle below, so that we see two plates moving towards one another and one being annihilated; finally, in conservative margins, plates are neither created nor destroyed but just slide past each other.

Although individual plates may be made up of continental or oceanic crust, or both, the constructive and destructive margins can only occur in association with oceanic crust. Continental material, as far as we can tell, is today neither created nor destroyed by the processes of plate tectonics. This is an important point. It is fairly obvious from direct observations of ocean ridges that only the thin oceanic crust is being created, but what would happen if a continental mass were carried up to a deep trench on a plate which was being destroyed at that margin? It seems that although the continental crust may be carried downwards a little way, its thick, low-density material (only about 2.85 gm per cm^3 compared with the 3.35 gm per cm^3 of the underlying mantle) cannot be absorbed by the trench, and the destructive margin soon ceases to operate. So the whole pattern of tectonic activity is a variable feature of the Earth. As we have already seen, it is extremely unlikely that the size of our planet is changing much today, or has changed for hundreds of millions of years. So oceanic crust must be being created at spreading ridges at just the same rate that it is destroyed in the deep trenches; if a deep trench gets blocked up with continental crust, a spreading process somewhere else in the world must slow down or stop in compensation. Indeed, any change in the velocity of a plate's motion (either its speed or its direction) must be counterbalanced by a corresponding change somewhere else. One dramatic consequence of this is that, just as we can point to the Red Sea as a newly active spreading ridge, we can locate regions of the Earth where spreading activity has stopped, recently by geological standards. Western North America, for example, has overrun the northern part of the Pacific spreading-ridge system, which is now extinct, although traces of its dying activity can still be identified. This is a particularly interesting special case, which has produced the San Andreas Fault system of California, where the spreading-ridge system of the South Pacific now runs into the North American continental mass; we shall look at the details of the unusual geophysical effects which result in detail in Chapter Seven, together with the equally interesting situation of the Great Glen Fault in Scotland, which once marked the site of a conservative margin.

To understand the overall picture of plate tectonics, it is important to bear in mind that the plates are moving not over a flat plane, but over the almost spherical surface of the Earth. The details can be found in the more specialized geophysics books mentioned in the bibliography; the vital point, however, is that the correct spherical calculations explain details of the processes operating at plate margins better than any oversimplified flat-plane model calculations. In general, the constraints which the geometry of the situation imposes are less restrictive than the physical laws which seem to govern plate tectonic activity here on Earth. For example, there is no geometrical reason why spreading should occur evenly on either side of a constructive margin; we can imagine spreading on one side of the ridge only, or spreading from both sides but at different rates on each side. However, this is not in line with what we find in practice.

There are, nevertheless, plenty of pretty complicated

situations, such as the tectonic processes occurring at triple functions, where three plates meet, or a spreading ridge which meets a destructive trench at right angles. An individual plate can be any shape, and can be bounded by any combination of constructive, destructive or conservative margins, as long as the whole global picture adds up to produce a constant amount of crust at any time. The African plate, bounded on the west by the Atlantic Ridge and to the east by the Indian Ocean Ridge, is growing larger on both sides; in the North Pacific, the largest plate in the world today is no longer growing at the eastern edge – a conservative margin along the coast of North America – but is being destroyed in the deep trenches of the West Pacific, so the plate is shrinking. Since the Pacific plate contains no continental crust, it is quite possible that it will be destroyed altogether and that some time in the future North America and Eurasia will collide to form a new supercontinent. By then, the Atlantic may have widened to the size of the Pacific today, and after the collision new spreading processes will break up the supercontinent and start the whole process over again.

Such collisions, rearrangements and break ups seem to have occurred several times in the history of the Earth. When continents collide, mountain chains are built up (similarly when subcontinents collide: the Himalayas are the direct result of India plowing northwards into Eurasia relatively recently), which provide a guide to the tectonic processes of previous cycles of continental drift. Mountains also form at the edges of continents involved in tectonic processes; as South America moves eastwards, oceanic crust slides into a deep trench along the Pacific coastline while mountains are built up above, and as we already know the mountains of Cyprus are being squeezed upwards as the African and European continents move together. Indeed, the example of what is happening today in Cyprus provides some of the most compelling evidence that tectonic processes are nothing new to the Earth. Mountains in Newfoundland contain just the same kind of rocks as those of Cyprus, including deposits of copper; presumably, they formed in the same way. But those Canadian mountains are 500 million years old, and the breakup of the supercontinent of Pangea (which included Laurasia in the north and Gondwanaland in the south) ties in roughly with the beginning of the spreading of the Atlantic Ocean floor, about 200 million years ago. So, at a time even longer before Pangea broke up than the time since then, tectonic processes exactly like those in the Mediterranean today were taking place. The rocks which now form the Newfoundland mountains were then an old sea floor, being squeezed into extinction as two continents moved together.

It is, then, not strictly true to say that the amount of continental crust on the surface of the Earth is unchanging. Sometimes, new mountains are created when sediments are scoured up, as in the Andes at the leading edge of a plate moving westwards; at other times and in other places blocks of oceanic crust can be squeezed upwards to become part of the continental mass when collisions between plates carrying continents occur. In such collisions continental crust can also be destroyed to a limited extent; for example, as India drives northwards into Eurasia, the Indian subcontinent is subducted under the Himalayas, the mountains being raised up in the process. But most subduction zones destroy oceanic crust, and very little new material is added to the continents even in a million years. However, there are a few places where tectonic activity occurs at a rate which is significant on a human timescale; for instance, to the west of the Straits of Gibralter, where the submerged parts of the Eurasian and African plates are pushing together in the Atlantic, there is a region of jumbled, rocky, sea-floor 'terrain' where features have been found to change within a few years and show signs of building towards the surface, squeezed by the collision of the two plates. Could this be the site of the next 'Cyprus', just as Newfoundland seems to represent an earlier similar event?

In spite of this evidence, and similar data from other continents, some geophysicists still argue that the breakup of Pangea was a unique event in Earth history and that the supercontinent itself was not the chance creation of drifting and colliding continents. But then there are also geophysicists who do not accept the idea of plate tectonics at all, and remain firmly convinced that the continents do not move about the globe. It is not entirely unfair to group the two together, since those geophysicists who have been persuaded – against what they see as their own better judgment – that drift has occurred for the past 200 million years tend to cling on to as much of their old training and beliefs as possible by rejecting the notion that drift occurred before the breakup of Pangea. That's human nature; but if the breakup of Pangea was a unique event, how do you account for the fact that the Pacific sea floor is nowhere as old as the rocks of the oldest continents? Surely this evidence that even the vast Pacific has been created in the past few hundred million years can best be explained by regarding tectonic activity as a continuing process since the Earth's surface reached the stage of having distinct oceans and continents – and maybe since before that time.

We are still some way from making any use of this extension of plate tectonics into the period before Pangea. As yet, we can use the presence of copper in Newfoundland as evidence that tectonic processes went on 500 million years ago, but we don't understand these ancient tectonic events well enough to predict where other deposits of copper might be found, except in the vaguest terms. But this need not always be the case; it is quite possible that further research and a better understanding of both present and previous tectonic activity will enable us to track down older deposits of this kind. Geophysical research does not come cheap by everyday standards; drilling holes in the ocean bed, for example, is not something to be undertaken lightly. But the economic value of such research is so obvious that surely it cannot be neglected.

The first step in reconstructing the continents that existed before Pangea is easy. If we accept that every mountain range has been formed by tectonic activity, and usually by the disappearance of an ocean, then all we have to do is 'tear along the dotted line of the mountain chain', as Nigel Calder puts it in his book *Restless Earth*,* to reveal the line of the ancient continental edges. And this is true even when

* B.B.C. Publications, London, 1972.

today the mountains are buried deep in the heart of a continent, as is the case with the Urals. These mountains formed about 225 million years ago, when Europe and Asia collided and stuck together – one of the last events in a previous phase of tectonic activity in the formation of Pangea.

The furthest back that we can push our knowledge of the changing face of the Earth at the moment is to about the time of the breakup of the supercontinent before Pangea, some 500 million years ago. This is particularly interesting because at about that time important changes were occurring in the evolution of plants and animals, with life moving out of the oceans for the first time. Then, the bulk of the land mass of the Earth was assembled in a kind of super-Gondwanaland around the South Pole. As the continental fragments moved apart, most of what is now Europe was part of 'Africa', with the rest of 'Europe' attached to 'North America'. About 420 million years ago, a collision between these two continents produced the mountains of Norway and Scotland, with activity in the Great Glen Fault (see Chapter Seven). After a further 100 million years, collisions with the rest of 'Africa' and with 'Asia' produced the mountain chains which now run east–west across Europe and north–south across Russia. Pangea was completed, but only to break up again after less than 50 million years, first into the two supercontinents Gondwanaland and Laurasia and then into more or less the continental shapes (but not the positions) that we see today.

Tectonic activity may well be a very old feature of the Earth. Before about 2,000 million years ago, geological evidence suggests that there were no large continents, and certainly today we can see that tectonic activity tends to enlarge the existing continents, as material from the sea floor is pushed up to help form new mountain ranges (as in the east of South America or in Cyprus), which then become permanent continental features. As yet we have no very satisfactory idea of what forces cause the tectonic activity; but we do know that they must be associated with fundamental processes going on in the interior of the Earth, which are discussed in the next chapter.

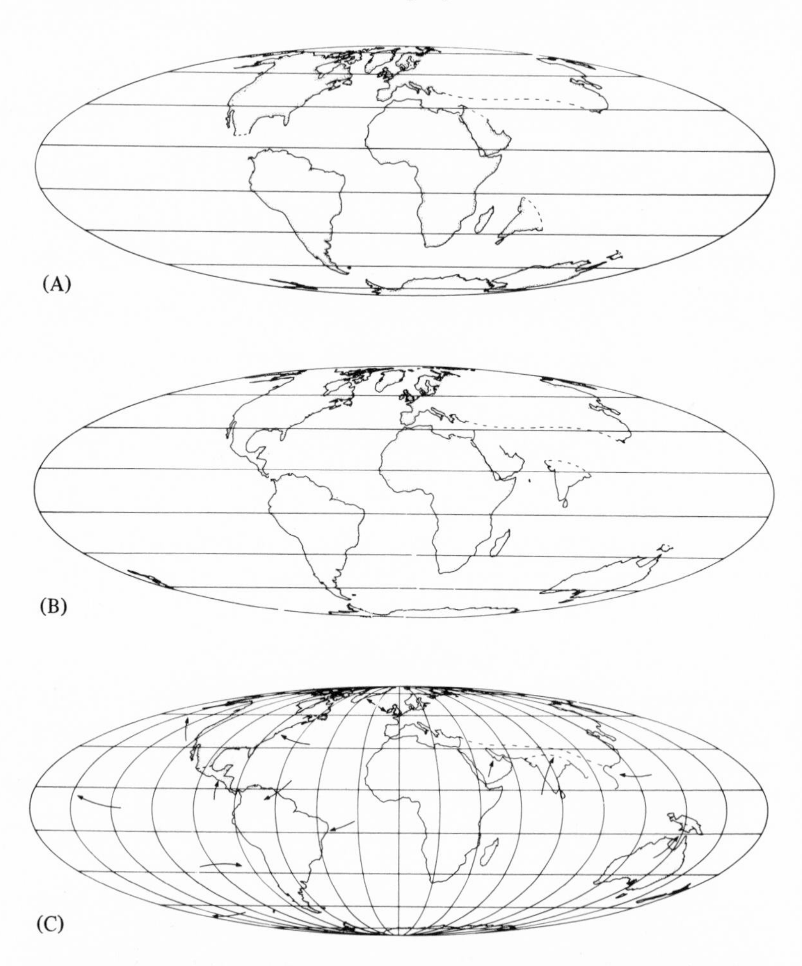

Fig.7 Reconstruction of the positions of the continents (A) 75 million years ago and (B) 35 millions years ago, compared with present positions (C) on the same projection. Arrows mark the directions of drift today. The African plate has been arbitrarily chosen as the 'stationary' standard against which the relative motions are measured. (Courtesy of Dr D. McKenzie, University of Cambridge)

Chapter Six

The Earth's Interior: What Drives Continental Drift?

The Earth's crust is made up of several plates, in constant motion relative to one another, with oceanic crust being both created and destroyed. But what is the driving force which pushes the plates around? Since we know that the interior of the Earth is hot, and that therefore some layers at least must be fluid, it is a simple step to suggest that convection is responsible for the plate motions. Most modern ideas elaborate on this theme in one way or another, but as it is far from straightforward to find out details of what is going on in the Earth's interior, the theories described in this chapter – although they are the most widely accepted modern views – may one day be proved incorrect. Remarkable progress has been made lately, especially in seismology, but the study of the Earth's interior is still by far the most difficult aspect of the new understanding of the Earth.

One of the great problems faced by geophysicists who study the movement of the continents is the vast scale – both in space and time – on which changes in the Earth's crust usually take place. However, an intriguing natural 'model' of plate tectonic processes has recently been found in the Kilauea volcano on the island of Hawaii. This reproduces on a small scale many of the features of the global tectonic system, and also provides a clue to the nature of the convective forces which drive it.

This miniature tectonic system grew up during an eruption of the volcano which lasted from May 1969 to November 1971 and it was monitored by members of the U.S. Geological Survey. During the eruption, a column of lava within the vent of the volcano became stagnant and formed a crusted surface. Previously, the column had been active, rising and falling over tens of metres and emitting gas, but the eruption of material from other parts of the volcano seems to have allowed it to become quiescent. In the stagnant condition, there was a steady convective circulation of lava in the column, which moved pieces of the glassy crust around in roughly the same way as the Earth's crust is moved about over the underlying material. Because of the convective movements in this volcanic lava the crust was repeatedly fractured; pieces were churned back into the molten material below in some places while in others fresh crust was generated as rising material cooled and solidified.

Of course, such a small-scale feature can only have a rough similarity to the tectonics of the whole Earth. But many of the parallels are remarkably exact. For example, crust is formed along lines, spreading outwards on either side as lava wells up from the column, while at other sites pieces of crust (plates) converge, with one plate being thrust under the other to remelt. In addition, more complex 'plate boundaries', including triple junctions where three plates meet, occur in the crust in the lava column in much the same way as they are believed to occur in the crust of the Earth. This discovery is more than an entertaining trick of nature, because such very close similarities with the models of global tectonics encourage great confidence not just in the overall tectonic model but also in the details of theories of how the tectonic processes operate at plate boundaries. And since the whole Kilauea miniature tectonic system is indeed powered by convection, there is very little room left to doubt that convection, in one form or another, drives the global tectonic machine. The question is, in just what form? And how can geophysicists find out anything directly about the parts of our planet which lie below the outer crust?

Although the answer to the first question is still not entirely clear, the choice seems to have been narrowed down to a few possibilities. However there is, as yet, only one real answer to the second question, since only by seismology can we find out anything reliable about the Earth's interior. This kind of seismology is far removed from detonating a few pounds of explosive in a borehole and recording the resulting sound waves in the rock. To study the deep Earth, you need much bigger shock waves – either natural earthquakes or, in the past few years, the disturbances caused by underground nuclear explosions. Seismologists have been studying the effect of the structure of the Earth on natural shocks from earthquakes for many years, but the science of seismology received a great boost in the 1960s when it became important to devise seismological techniques which could distinguish nuclear tests from natural earthquakes. With this political motivation the science received the funds which helped to develop the techniques which now tell us so much about the planet beneath our feet. Ironically, from the point of view of the politicians, the improved understanding of natural earthquake seismology which was a by-product of this has now led to a situation in which, under some circumstances, nuclear explosions can be set off in such a way that the shock waves they produce mimic those of natural earthquakes. We are still a long way, it seems, from having the know-how to enforce a comprehensive ban on underground nuclear testing. But that is another story.

The painstaking way in which seismologists have built up a picture of the Earth's interior is one of those many areas of science where quiet, methodical work provides the sound base for new theories and new flights of fancy. No progress could be made without such groundwork, but the story does not lend itself to dramatic interpretation and I shall only mention a few highlights before describing the overall structure of the Earth's interior as it is now understood; anyone who has a taste for the details of the story should turn to the chapter on 'The Fine Structure of the Earth's Interior' in the book *Planet Earth*.*

Some idea of the sophistication of modern seismological techniques is seen from studies of high-speed earthquake waves reflected from the Earth's core. In general, earthquake waves (whether natural or artificial) can travel from the source to detectors in other parts of the world either by a

* W.H. Freeman, Reading, 1975. Readings from *Scientific American*.

more or less direct route through the upper layers of the Earth or by being reflected from one or more of the boundaries between different regions of the interior. If that was all that happened, the seismic waves would be easy to interpret. But, like light travelling through different mediums, the sound waves from seismic events can be bent and their speed altered as they move through regions of different density. So it is necessary to know the time at which a seismic disturbance takes place, as well as the time when the different kinds of earthquake wave reach the receiver, in order to find out how fast the waves have travelled. With several detectors monitoring one earthquake and all the different waves produced by it some of the unknown factors can be eliminated, and today there are several large arrays of seismometers which are used almost in the same way as large radio telescope arrays are used to interpret radio noise from objects in the depths of space. The Large Aperture Seismic Array (LASA) in the U.S.A. consists of 525 linked seismometers grouped in 21 clusters which cover a total area 200 kilometres in diameter; it is small wonder that our understanding of the Earth's interior only really developed when such arrays became possible through the provision of 'defence' funds. The detail that emerges from studies with such giant arrays shows not only the division of the Earth's interior into different layers, but also the echoes from particular features near the surface, such as the broken-off pieces of old oceanic crust still sliding down into the depths under the Eurasian landmass, where the welding together of two former continents has now obliterated the ocean and trench system which once existed. When large underground nuclear tests are widely advertised in advance, like the Cannikin explosion of November 6th, 1971, they provide the best probes of all for studying details of the Earth's interior as the results can be specially monitored in all corners of the globe.

Cannikin was an underground test exploded by the United States in a deep hole on Amchitka Island, near Alaska, and the shock waves which it sent around the world were monitored by many seismic stations. In Australia, a team from the Australian National University, Canberra, made elaborate preparations and carried out some very successful observations. Three weeks before the explosion, they set up ten portable seismic stations between Canberra and Maralinga in the eastern part of Australia, which added to a network of four permanent stations run by the A.N.U. and Adelaide University to provide a straight line of detectors stretching for about 1,700 kilometres in a roughly north-west–south-east direction. By studying the well-documented explosion and arranging their chain so that each station in it was at the same distance from the test, the Australians were able to eliminate many of the variables which cause uncertainty in the interpretation of more usual seismic disturbances. The result of all this was that they were able to determine with great confidence that the signals recorded at the fourteen detectors arrived 1½ seconds earlier in the west than in the east, and this shows that there is distinct variation in the velocities of these seismic P waves in the mantle under different parts of Australia; the increase in velocity towards the west is particularly marked about half-way along the line of detectors, and this suggests that the region of mantle in which seismic velocities are low (the low velocity zone) thins out in that region, which is probably the edge of an old continent.

In a notable contrast to the scale of Cannikin, seismic equipment is now so refined that in one test reported in 1972 seismic waves from the explosion of only ten tons of T.N.T. in the North Sea were picked up as far away as Brazil and Australia. This depended on having some knowledge of the Earth's interior in advance, since the size of the charge and the depth at which it was set off in the sea were carefully calculated to optimize the chance of it being detected at distances. The success of this experiment by the U.K. Institute of Geological Sciences shows both that this understanding of the Earth's structure is correct and that such small explosions can now be used to find out details about the structure of the crust. One of the best aspects of this particular story is that the I.G.S. team did not warn anyone in advance about the explosion, but waited to see if their colleagues around the world could indeed detect it without any clues. As well as stations in Britain and the continent of

Europe, reports came in from the U.S.A., Canada, Australia, Africa and Brazil, with even the more distant stations able to estimate quite accurately the size of the explosion, even though the seismic waves were obviously quite weak by the time they had travelled such distances.

This technique, however, really applies to the Earth's crust, rather than revealing new secrets about the interior of our planet. Nevertheless, the deeper travelling seismic waves tell us that the Earth has several layers, from an inner solid core about 1,000 kilometres in radius out to the plates themselves. These layers are shown in Figure 8. It is particularly interesting to notice that the liquid core does not extend right to the middle, but has an inner solid core almost as big as our Moon, and that the surface layers of our planet are subdivided several times, so that the idea of a solid crust

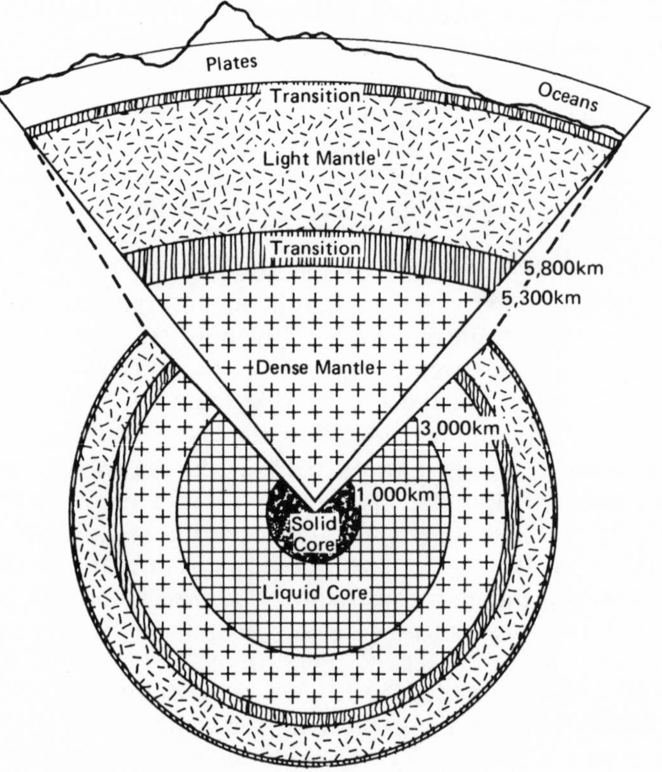

Fig.8 The layering of the Earth.

overlying the mantle is very much an oversimplification. In terms of finding out what drives the plate tectonic machine, it is worth looking a little closer at these surface layers.

The mantle itself is all solid but it is divided into two regions, separated by a transition zone, because at greater depths and pressures the crystal structure of the rock is rearranged so that the material is denser and takes up less space. Above the lighter part of the mantle there is another transition region, the low velocity zone which I have already mentioned, which seems to be made up of a mixture of molten rock and pieces of solid rock being recycled from the plates above. Above this lubricating layer the plates themselves, some 64 kilometres thick, carry the jigsaw puzzle pieces of the visible crust on their backs. The crust itself is only a few kilometres thick in oceanic regions, and some tens of kilometres thick in continental regions; the plates are a part of the upper mantle.

So even when we look at the details of the supposed convection processes which drive the plate motions, we are only dealing with a tiny fraction of the whole Earth. However, before we look in any more detail at that small volume which is of special interest in understanding the nature of our changing Earth and the formation and location of pockets of natural resources, we should perhaps consider another area of research which provides some clues and half-answers to the question of what lies beneath our feet. Although not as reliable as seismology, the study of the Earth's magnetic field has its own fascination.

The magnetic field of the Earth clearly originates somewhere (and somehow) in the interior, and the only widely accepted explanation is that it is generated by some form of dynamo action, set up by electric currents in the fluid part of the core. This is a very vague explanation, and because so little is known about conditions deep inside the Earth (even with the aid of modern seismic techniques) acceptance of the general idea still leaves plenty of room for speculation. Indeed, it has not proved simple to produce a theory which shows that dynamo action of this kind will work at all inside the Earth. Given all these difficulties, it is a real triumph of modern Earth science that some researchers have not only produced a reasonable theory of the dynamo effect, but have also suggested how the dynamo can reverse from time to time, bringing the magnetic reversals which provide the 'stripes' on the sea floor which were so important in persuading geophysicists of the reality of continental drift and sea-floor spreading.

The model first took shape as a mathematical description of what might be happening inside the Earth. The theory describes the dynamo effect in terms of a conducting, stationary sphere in which are two smaller, rotating spheres. In its simple form this is only a very rough approximation to conditions in the Earth, but the success of the mathematical predictions encouraged a team at the University of Newcastle upon Tyne to build an equivalent physical model in their laboratory. In this, the spheres were replaced by cylinders made of steel, which rotate inside holes in a steel block. The cylinders are free to rotate and to be tilted relative to one another, all the while being kept in electric contact with the stationary block by a film of mercury. And under certain conditions the two rotating cylinders do produce a dynamo-generated magnetic field.

Although this model only mimics some major features of the Earth, it produces a field which oscillates over timescales of 20 to 40 seconds, mimicking the movement of the Earth's magnetic poles relative to the geographic poles; under the right conditions it also produces complete reversals of the generated magnetic field, reminiscent of the geomagnetic reversals. The study of these processes still has a long way to go, but it looks as if the researchers may be on the right path. Inside the real Earth magnetic reversals may be triggered by changes in the pattern of convection in the fluid core, caused by small changes in the production of heat at different sites inside the Earth.

The pattern of the Earth's magnetism may also tell us something about the 'roughness' of the boundary between the mantle and the fluid core. Irregularities on the bottom of the mantle will disturb the fluid core, and this will produce changes in the flow which is producing the magnetic field. Some studies in the late 1960s and early 1970s have produced evidence that the pattern of the Earth's gravitational field, which shows up irregularities in the thickness of the mantle, can indeed be related to the magnetic pattern. But this kind of work is several steps harder than attempting to explain why the Earth has a north and south magnetic pole at all, and it is hardly surprising that some of the claims made by geophysicists investigating this kind of detail have been strongly opposed by their colleagues. So it seems best to leave the murky depths, where the interface between core and mantle certainly generates controversy – whether or not it generates magnetic disturbances – and come back to the upper layers to take stock of the best interpretations of the convective mechanism which moves the plates, carrying the continents, about on the face of the Earth.

There are four basic mechanisms which could be moving the plates, two of which do not really involve convection directly. It seems quite likely that in the real world all four processes operate, with different relative importance in different parts of the globe; but it is easier to see what is going on by looking at each of them separately. First, there are the simple gravitational effects caused by the weight of the plates themselves. At ocean ridges, material is built up to a greater thickness than usual, and with the lubricating layer just below it seems quite plausible that the new crust would slide away downhill on either side of these ridges. When crust is being destroyed at a deep ocean ridge the layer that is diving down into the mantle to melt and be recycled persists as an intact sheet for a long time, and down well below the crust of the overlying continent, as the study of seismic reflections shows. Once such a process is under way, the tug of this older oceanic crust sliding down into the Earth's interior is likely to encourage the general spreading of the sea-floor crust outwards from the ocean ridges. Both these gravity, or weight, effects can contribute to the changing face of the Earth once tectonic activity begins, and they must have some effect in the world today, even if this is only important close to the ridges and trenches. But convection remains the only serious candidate for the basic driving force which started the whole business off and still keeps the global tectonic machine in motion.

When any fluid is heated there is a natural tendency for the hotter material to expand, which makes it less dense so that it rises towards the surface. At the surface, it loses heat and drifts sideways, eventually cooling, becoming more dense again, and sinking downwards. The whole process then repeats, with the fluid material moving round and round in convective 'cells'. It is easy enough to visualize this happening in a pan of water on the stove, and you may have seen the decorative lights filled with coloured oil which rises and falls in convective patterns due to the heat from the bulb in the base of the lamp. But it is hard to imagine rock acting in this way, even though the laws of physics tell us that it must. The convection of rocky material is, of course, very slow by human timescales, and it is also different from the convection in a pan of water (or a decorative lamp) because the rock is not heated from below but by the radioactive decays which are constantly going on throughout the bulk of the fluid material. But this turns out to be just what is required to explain why the convective cells of the Earth's upper layers are so large.

When material is heated from within in this way, it does not break up into many small, square convection cells (which is what happens to your pan of water on the stove) but forms very elongated oblong cells, much longer than they are deep. The depth of this convecting layer at the surface of the Earth is only a few hundred miles, but the sideways extent of a convective cell, from spreading ridge to ocean trench, can be many thousands of miles. As yet, the picture is still rather vague. A map of plate boundaries (see Figure 4, page 60) does not show any obvious neat way of dividing up the Earth into convective cells, but as usual the neat picture of what happens in the laboratory is bound to be distorted by the complications of the forces which change the face of an entire planet. One particular possibility both emphasizes the difficulty of untangling these complexities and also provides a possible new tool for investigating how plate motions have changed in the past.

This is the concept of 'hot spots', which has been suggested as the fundamental driving force of tectonic motions, but which to my mind looks better as an additional, complicating factor acting as well as the overall convective flow. In this picture, there are perhaps a score of sites around the globe where hot material rises in a narrow column or pipe from deep in the Earth's interior. The powerful localized effect of this rather different form of convection could contribute to the jostling of the plates of the Earth's crust; but even more interestingly, the presence of hot spots can account for the formation of such systems as the chain of Hawaiian islands, a volcanic chain produced in the middle of the largest oceanic plate of all, and well away from the regions of geological activity which we expect to find at the edges of plates.

Hawaii itself is the youngest of these islands, and is still volcanically active; it lies at the south-east end of the chain. Moving away to the north-west, the other islands in the chain are no longer active, although they too formed as volcanic features, and intriguingly the islands further from Hawaii are older. The chain has emerged as volcanoes have popped up, one by one, along a straight line which has just one bend in it, where the oldest islands of all turn away on to a more northerly line, instead of continuing the north-west trend of the younger part of the chain. In terms of hot spots, the explanation of this is simple. Today, as we know from other evidence, the Pacific plate is moving north-west. So, if there is a hot spot under the middle of what is now the Pacific plate, it may have punched a series of holes through the thin oceanic crust which is moving above it. Each hole will have formed into a volcanic island, and as older islands are carried away by the plate to the north-west new islands appear behind them at the site of the hot spot.

What about the bend? The simple explanation is that about 40 million years ago the Pacific plate changed from a more northerly movement to the north-westerly drift we see today, and so the older islands were formed during a slightly different phase of tectonic drift. At least two other island chains in the Pacific show the same kind of pattern as the Hawaiian chain, including the bend in the chain equivalent to a date 40 million years in the past, and if this hot spot idea is correct, between them the chains provide a record of the movements of the Pacific plate over about 100 million years. It is interesting that about 30 to 40 million years ago the San Andreas system in California (of which more later) began to form as North America overran the line of the northward extension of the Pacific Ocean spreading ridge; could it be that this collision between plates played a part in changing the direction of drift of the Pacific plate, and produced the bend in the Hawaiian and other chains? This is speculation – but then, the whole of modern geophysics is built on what used to be speculation!

If the Hawaiian chain provides a very clear-cut example of the effects which are supposed to be produced by hot spots, in other parts of the globe the situation is more confused. On the hot spot theory, Iceland is such a site, and the old volcanic islands of Scotland were made by the same hot spot that now produces the activity in Iceland. But Iceland, unlike Hawaii, can be explained quite satisfactorily as the natural product of a spreading ridge, and it lies on a boundary between plates, just where we expect to see volcanic activity even without invoking hot spots. Again, though, the presence of a hot spot under the U.S.A. could neatly explain the geological activity of the Yellowstone region. And so the argument goes back and forth. Today, the controversy about whether or not hot spots really do exist is one of the most furious in geophysics, and even some of the pioneers of plate tectonics, including Dan McKenzie, find the concept just too much to swallow. I personally find no difficulty at all in accepting the reality of hot spots; indeed, they seem just the sort of complication that would be likely to stir up the plate motions and stop the formation of nice, neat convective cells that could be easily distinguished from geological maps.

However, whatever the details, the energy involved in moving the continents around by the convection produced by heat inside the Earth is impressive. We are now stuck in an energy crisis which seems likely to persist for twenty years or more, even if the hopes of tapping nuclear fusion energy to provide unlimited clean power are realized; so is there any prospect of tapping some of this vast reservoir of geothermal power to tide us over the next few decades? The answer seems to be 'yes' – but a cautious yes.

We have known about geothermal power for a long time – the energy in geysers and hot springs is plain for all to see. However, it is only in the past few years that geophysicists have come to realize that the relatively few locations around the globe where the heat of the Earth's interior bursts right through to the surface in this way are greatly outnumbered by regions where strata containing steam and hot water lie fairly close to the surface, ready to be tapped by drilling. The heat of these geothermal reservoirs comes from poorly understood variations in the underlying convection, which in some places seems to push bulges of magma up into the bottom of the crust – mini hot spots, if you like. If the rocks above such a magma dome are porous or fissured and contain water, the water will soon become superheated (we are still talking about depths of five or six miles – eight or nine kilometres – below the surface, where the pressure is so great that even water at 500° F will be liquid). If the superheated water escapes to the surface through a crack in the rock, it will boil and blast outwards as steam – and if there are no natural cracks, then a man-made well drilled into the porous layer will do just as well.

Already more than a million kilowatts of energy are produced from geothermal sources around the world, and by 1980 it is likely that this figure will have risen to at least five million kilowatts. In one area of northern California alone (the Geysers field) there is a potential capacity estimated at three million kilowatts, and in the Imperial Valley of southern California perhaps as much as 20 million kilowatts of geothermal potential is waiting to be tapped. In Italy, an electricity generating plant has been powered by steam from the Lardello field since 1904, but the growth in the use of this supply of clean energy has only started to pick up recently, as the cost of other energy supplies has increased.

Geothermal reservoirs occur naturally in three types: dry steam fields, where no water escapes; wet steam fields, where the steam emerges mixed with hot water; and fields where the water is not hot enough to boil and rises as hot springs. Even the hot water is useful – perhaps for central heating – but the greatest power, of course, comes from the steam, which can be used to drive turbines and produce electricity directly. The low pressure of the natural steam limits individual generators to about 50 to 60 megawatts, small by oil-fired standards, but the advantages of geothermal power make up for this.

In addition to the power itself, the steam can be used as a source of freshwater, and the water found in wet steam fields and hot springs is often rich in minerals, like the brine pools of the Red Sea. In all-round terms, a wet steam field offers the greatest potential, since it can produce energy, hot water for heating systems and mineral-rich brine which can be evaporated to provide raw materials for industry. It sounds almost ideal, especially since the development of a geothermal field causes very little pollution compared with conventional energy sources. So why only a *cautious* 'yes' in response to the question of whether geothermal power is what we need to tide us over the next few decades?

Quite simply, the problem is that geothermal power only provides a finite resource, just as reserves of oil and natural gas are finite, so the development of geothermal fields can only be a short-term prospect even in human terms. The best estimates seem to be that a typical field could last for thirty years, with proper management and development, and while that is long enough to help considerably in the present crisis it is also short enough for us, rather than our children, to be worrying about what happens afterwards. Of course, as exploration and development open up new geothermal fields the available power will gradually be extracted, taking the world as a whole, over much longer than thirty years. But geothermal power cannot provide the ultimate, long-term solution to the energy crisis.

One of the factors involved in good management of a natural field is to pump water back down into the hot rock to replace the steam extracted. This water will be heated in turn and vented back to the surface as steam, but it is just not possible to supply economically the enormous quantities of water that would be needed to replace completely the amount removed – at the Wairakei field in New Zealand this is some 70 million tons of water a year. And, of course, if we are to tap geothermal reservoirs which have not broken through to the surface under their own steam a lot of drilling is involved, which can cost hundreds of dollars or pounds for every metre of every well drilled. If we consider the cost of the piping and all the usual paraphernalia of power stations, geothermal power no longer looks quite so cheap in absolute terms after all. It is this basic cost – the amount of effort in terms of both money and energy that has to be put in before power can be extracted – that makes some of the most adventurous ideas for tapping geothermal power look a little bit silly.

According to some prophets, we need not worry too much about locating regions where natural reservoirs of superheated water lie conveniently close to the surface, because there are so many regions where hot rock without any water content lies a few miles deep. So, the argument runs, all we have to do is drill deep wells in these regions, pump water down one and wait for steam to come blasting out of the neighbouring holes. The idea sounds even better when its advocates point out that the cold water pouring down on to the hot rock will crack it and open up convenient fissures for the water and steam to spread along. However, they are unlikely to make any profits out of such schemes. I have already mentioned the high cost of drilling the wells, which rapidly moves towards a million dollars *per well* when we start talking about depths of ten kilometres or more, and several wells would be needed to create an artificial steam field. In addition, the water has to come from somewhere, and by and large water is in short supply on our planet. Seawater would have to be raised uphill at least a little way even for drillings near the coast, and most rivers in the countries that need power so desperately are already being used for one purpose or another by industry. Imagine the outcry if either the river Thames or the Hudson river were diverted down a hole in the ground to provide steam for a geothermal power station! That's an exaggerated example, of course, but it sometimes seems that the visionaries who put forward ideas for solving man's energy needs at a stroke have not thought out their plans in full detail.

You may argue, though, that as other fuels are so scarce any cost in developing new resources will be economic. This is true in money terms, but look at it another way. The drilling of wells is expensive primarily because it uses up a great deal of energy; providing water for this kind of geothermal power station would also involve much energy, and it is extremely doubtful whether the energy extracted from the field over its useful lifetime would, in fact, be more than the energy that had to be put into the field to get it going in the first place. This is not a unique situation. According to some calculations, the amount of energy involved in developing the oil reserves of the North Sea (in industry, building the drilling rigs; transporting the rigs out to sea; transporting the oil back to land; refining the oil; etc.) may not be much less than the amount of energy (in the form of oil) ultimately extracted. So very careful planning is necessary if we are not to end up in the same situation as the man who lost a shilling and found a sixpence. Geothermal power really is the best short-term answer to the energy crisis (except for the even simpler solution of switching off everything that uses energy), but in the long run we will have to tap the unlimited power of nuclear fusion, either in man-made reactors here on Earth or by efficient use of the energy we receive from the great natural fusion reactor of the Sun. Meanwhile, and long after we have all disappeared, the convective forces operating in the hot rock of the Earth's interior will continue to move the plates of the crust about in an ever-changing jigsaw puzzle pattern.

Chapter Seven
The San Andreas Fault and The Great Glen: Two Case Histories

Now that we have examined at least the outlines of the overall world picture, we can see very clearly just how the slow, remorseless processes of geophysics relate to human experiences by looking in detail at two parts of the world which have been involved in the same kind of change. In Scotland, the geophysical processes which formed the Great Glen, which contains Loch Ness, home of the notorious 'monster', have just about run their course; in California, the activity along the San Andreas Fault represents an early stage in exactly the same kind of activity. So by looking at these two case histories we can see both the beginning and the end of the effects of lateral movement along a fault line. They make a particularly good choice for a detailed study, since horizontal movements of blocks of crust around the surface of the Earth are, of course, a key feature of the new geophysics; before plate tectonics became established, the conventional view was that crust could perhaps move up and down, but it certainly could not move sideways.

The San Andreas Fault deserves a whole book to itself;* however, deliberately leaving aside much of the detail, its overall nature can be summarized fairly concisely. The most important point is that the coastal region of California, to the west of the fault, and the Baja California peninsula do not belong to the North American continent at all. They are part of the Pacific plate, and the line of the San Andreas marks the boundary between that plate and the North American plate. Along this part of the plate boundary, the

* See Robert Iacopi, *Earthquake Country* (Lane Books, California, 1975).

margin is neither constructive nor destructive; the two plates rub shoulders, with the Pacific plate twisting slightly anti-clockwise, so that, viewed from North America, the edge of the plate is moving from left to right (northwards). Of course, from the point of view of someone in Monterey, or the near-by Californian coast, it is the rest of America which seems to be moving from left to right, looking to the east across the fault, in a southward direction. Both points of view are justified; from whichever side you look you see the other plate drifting to the right, so the San Andreas is termed a right lateral fault.

The drift is not altogether smooth, however, because of friction where the two plates rub together, and along much of its length the right lateral movement of the San Andreas proceeds as a series of jerks. Tension builds up until the friction can be overcome, then the plates slip suddenly by as much as several tens of metres, then there is no movement again for perhaps a hundred years as strain builds up. The jerks which release the strain are, of course, earthquakes; the longer the quiet interval between jerks the worse the next earthquake is, and at present the southern part of the fault, including the Los Angeles area, has been ominously quiet since 1857. But, important though the jerky nature of the slip is to people who live in the area and suffer the consequences in the form of sometimes very severe earthquakes, in terms of the changing Earth the two plates slide past one another pretty smoothly. What's a hundred years, after all, on the timescale of the chart at the end of Chapter One?

The situation when two or more plates interact seems complicated because there is no fixed reference point on the surface of the Earth against which we can measure their movements. All of the Earth's surface is made up of plates, and all of them are moving in a changing pattern. When we look at the Atlantic Ocean, say, it is convenient to mention that the ocean is growing wider as material spreads out from the Mid-Atlantic Ridge, but that simple picture obscures the fact that the continents on either side of the ocean are involved in other tectonic processes as well. As it happens, the Atlantic Ocean is pretty symmetrical, and that is why it is such a convenient example of the sea-floor spreading process. But the Pacific is not symmetrical, and this is because the tectonic processes operating on the eastern margin (the North American coast) have not, during recent geological time, exactly mirrored the processes going on at the western edge (the Asian coast). This complicates the picture, but from studies of the magnetic anomaly patterns of the Pacific Ocean floor, and using the other new tools of their trade, geophysicists can give us a broad outline of how the Pacific got into its present state.

Not too long ago, by geological standards, the North Pacific was much wider and the spreading ridge associated with the ocean basin ran right up to the north. Today, the spreading region is clearly seen in the southern ocean, but it stops short where it runs like a knife into the North American continent – at the site of the Gulf of California (see Figure 4). Traces of the activity of the former spreading ridge can be found from seismic studies of the region underlying North America today; clearly, what has happened is that the continent has overrun the former spreading ridge, crushing it out of existence like a steamroller flattening a crack in the tarmac. Superficially, this is puzzling. If there was a *spreading* ridge in the North Pacific, pushing out new oceanic crust on either side, how could the continent ever overrun it?

The answer is simple. Spreading in the Atlantic Ocean has been occurring much faster than in the Pacific, so the American continent was being pushed westwards faster than the Pacific ridge could create new crust. For a long time, the Pacific Ocean crust was disappearing into a deep trench down the west coast of America, being gobbled up faster than it was being produced at the spreading ridge. It is rather as if an escalator, with moving floor being produced steadily at the bottom, was being 'destroyed' twice as fast at the top, so that the top end of the machine advanced steadily downwards. When the constructive and destructive zones meet, the result in this case is mutual annihilation – we have been left with a plate boundary where material is neither being created nor destroyed. Of course, Pacific Ocean material is still vanishing under Asia at the deep trenches along the western edge of the plate, and with the loss of the spreading activity the North Pacific is still shrinking. Eventually, this oceanic plate will be completely gone and a collision between Asia and North America will produce new mountains, welding together a new supercontinent as a prelude to the next phase of continental drift.

But what about the San Andreas Fault? This feature is very important (in human terms) because it marks the end of the spreading ridge which still remains to the south; it formed through the geographical accident of the North American continent swinging away south-eastwards at the 'bottom' of California, and tapering down into Mexico, so that it was the northern part of the spreading which was gobbled up first as the Americas moved westwards over the Pacific. Much of the early development of California was along lines common at destructive plate boundaries; the material diving down into the deep trench and under the continent must have helped to build up the inland mountain ranges – the Rockies – as the continent crumpled a little and as new material from the melting oceanic crust was added to the bottom of the continental crust near the edge. Rivers flowing off the young mountains carried sediments into the deep trench system just off the coast to the west, and when the continent eventually overran the spreading ridge this sedimentary material was scraped up on to what used to be the western side of the ridge, rising above sea level to form the coastal ranges. The broken-off western part of the former ocean floor is now sinking down into the Earth below the western U.S.A., leaving the coastal ranges, riding on the remaining piece of the Pacific Ocean crust and being carried northwards by the general twisting movement of what is now the Pacific plate.

This picture ties in well with the geological evidence from the north of California, but the picture to the south is a little different. This seems to be because the spreading of the sea floor does not occur absolutely uniformly, but along a series of parallel strips, rather like many conveyor belts running alongside each other; the strip just to the south carried its share of the spreading activity in such a way that the ridge struck obliquely into 'Mexico', remaining active for a while and slivering off a slice of the continent, which we now see as Baja California. The northward movement of the Pacific plate has now jammed this sliver of continent hard up against the Los Angeles region, creating the San Bernardino mountains. Today Baja California is pivoting against the mountains, so that the Gulf of California is still opening up, and further north California is beginning to be torn apart along the line of the San Andreas Fault. This tearing is the origin of the deep depression of the Salton Sea.

All this lateral movement is impressive. It is about 30 million years since the ridge and trench systems began to annihilate one another along the line of what is now the San Andreas; at that time, 'San Francisco' was some 1,000 kilometres further south and the right lateral movement along the fault continues today at a rate which averages out at six centimetres a year. The notorious earthquake activity of the region is, therefore, easily explained in tectonic terms. But the destruction of the ridge and trench system has had repercussions far outside the small local region of California. We saw earlier how the drift of the Pacific plate over the hot spot which has produced the chain of Hawaiian volcanoes changed abruptly a few tens of millions of years ago. With similar evidence coming from other volcanic chains, it looks as if the Pacific plate received a nudge which knocked its overall drift into a more north-westerly trend at just about the time that North America overran the spreading ridge. This is hardly coincidental, and it looks as if the jolt from the continent ploughing over the ridge did indeed knock the Pacific plate sideways. Once again, we see how closely the plates interact, and how important it is to take account of the broader picture of global movement even when looking at details.

The details of the San Andreas Fault are impressive on a human timescale. For 700 miles, from the Mexican border to the coast at Mendocino, the rift is marked by an almost continuous line of clearly visible features. These features vary considerably, of course, and it is perhaps because of this that even very few native Californians appreciate just how much of the topography of their state is the direct result of activity along the San Andreas Fault. However, two features make it easy to plot the trail of the fault on the ground: first, it is essentially a straight line, so if you do lose track of it, it is easy to look ahead in the right place to find the next obvious surface feature; secondly, by its lateral movement, the activity along the fault either brings into juxtaposition geographical features that do not belong together or slices features apart to produce obvious curiosities such as steep cliffs rising sharply from flat valleys. From the air, the scar of the fault can generally be seen alongside the Coast Ranges on the flight from San Francisco to Los Angeles, but the scope of the feature is even more impressive when it is traced on the ground.

Robert Iacopi gives a very full description of the main features to look out for in his book *Earthquake Country*, which is essential reading for anyone with an interest in geophysics who visits California. Here, therefore, I shall only mention a few of the principal signs of fault activity. The character of the fault varies considerably; in broad terms it is made up of five sections, three showing more or less continuous 'creep' activity as the two plates rub past each other, with the other two sections locking in place for scores of years at a time and moving in a series of sudden jerks – major earthquakes such as the San Francisco disaster of 1906. From Cape Mendocino north under the sea the fault is active; from the Cape south to San Francisco and a little beyond marks the line of the 1906 break, which overlaps slightly with the second region of creep activity, between San Francisco and Parkfield. The line of a great earthquake in 1857, from Parkfield to San Bernardino just east of Los Angeles, marks the second region of jerky movement, and from San Bernardino south across the Mexican border and into the Gulf of California the fault is again in a state of more steady creep. Where the fault creeps it is relatively harmless, since the steady movement releases tension almost as soon as it begins to build up, with the plates slipping at about five centimetres a year. It is the regions which are quiet today that are paradoxically more dangerous; along the line of the 1857 break, for example, accumulated strain equivalent to more than 120 years of slip has built up, and an earthquake producing sudden movement of about six metres seems inevitable in the not too distant future. Yet still people live and work near the fault, and have even built dams, hospitals and schools in the danger zone.

One of the best natural markers of recent creep activity is provided by the curved line of a stream crossing the fault. Because of the gradual movement, the streams tend to keep to their carved channels without breaking out into new banks as the eastern and western halves of the stream are pulled apart sideways. The result is a series of stream channels with sharp curves where they cross the fault; in one study of the central active region south of San Francisco 130 streams with channels displaced in this way were found along a 110 kilometre stretch of the fault, so they are not too difficult to find – especially since, in some cases, the offset of the two halves of the stream is more than 300 metres! A similar effect can be seen where some of the works of man spread across active parts of the fault; for instance, in the city of Hollister, inland from Monterey Bay, sidewalks, pipes and walls crossing the fault line have all been bent by the persistent right lateral creep. Roads are also affected where they cross the fault line, both by creep and, of course, by larger disturbances. A major quake in the fortunately sparsely populated region near Parkfield in 1966 produced a clear break across Highway 46, which could be easily measured because the fracture ran right through one of the white lines of the road markings. On the day of the quake (June 22nd), the offset produced was about five centimetres, but by August 4th creep following the quake had stretched the offset to more than twelve centimetres. This is a particularly interesting region of the fault, since the boundary between steady creep to the north and ominous calm to the south is not clear-cut, and peculiarities such as this sudden jump followed by relatively rapid creep combine features of both kinds of activity.

I have driven along roads near part of the San Andreas Fault, from the region of San Jose south to the mission of San Juan Bautista, which happens to lie at the southern end of the line of the 1906 rupture and covers the overlap with the region of creep to the south. Near Los Gatos, the mark of the 1906 quake is clear at the abandoned railway station at Wrights, where the five foot (1.5 metres) lateral movement shifted one end of the railway tunnel sideways, rendering it rather inadequate as a transport route. From the road to San Juan Bautista the line of the fault can be seen very clearly in places, especially where the road runs alongside the river and railway in a valley along the fault line. The cutting for the road in Chittenden pass reveals shale strata on the east and granite on the west – different rocks riding on different plates of the Earth's crust. Although it is clearly convenient for the roadbuilders to follow this line, it does produce a slightly queasy feeling when you know enough geophysics to realize that two wheels of the car you are riding in are on the North American plate and the other two are on the Pacific plate; even if events like the 1906 quake do only occur at intervals of 100 years or more, it doesn't feel like the safest road in the world.

Another place where builders have taken advantage of the fault topography is at San Juan Bautista, where the scarp marking the old line of the fault forms the foundation for the grandstand of the local rodeo. The Mission itself is built at the top of the escarpment; it suffered severely both in the 1906 quake and in a less well-documented but seemingly equally severe series of tremors at the beginning of the nineteenth century.

A little further to the south the influence of the great 1906 quake ends completely, but one of the most famous examples of creep at work is to be found in a vineyard near Hollister. This example came to light in 1956, when it was noticed that the reinforced concrete walls and floor slabs of a warehouse at the vineyard (now owned by Almadén Vineyards, and producing a very pleasant 'riesling', among other delights) were cracked and collapsing; it turns out that an active branch of the fault runs right underneath the building, tearing it apart at a rate of about one centimetre a year. In this region, several roughly parallel lines of the fault are involved in the right lateral slip, which is why the movement is less on this one branch than the five centimetres or so a year appropriate for the overall relative movement of the two plates. As of 1974, when I visited the site, the building (constructed in 1939 to replace an earlier warehouse which fell down, for reasons which are now obvious) was still standing but did not look the sort of structure I would like to work in; perhaps it will have gone the way of its predecessor before you can see it, but if it has there is further evidence of the activity of the fault where the same branch crosses a concrete drainage ditch south of the main building. This ditch shows signs of repeated patching, but it is still broken and displaced at the fault line; it is also clearly a notable tourist attraction for two other groups of sightseers appeared, cameras at the ready, during the few minutes I was there.

Fig.9 The author examining the broken drainage ditch at the Almadén Vineyard. (Reproduced by permission of John Faulkner.)

We can see such obvious details in the San Andreas system because it is still a young and active feature. But if anyone should think that geophysicists have now got all the answers, and that only a few mundane details remain to be painted into their world picture, the Great Glen in Scotland provides a salutary reminder of how little detail we can really claim to understand. The Great Glen Fault was once very like the San Andreas today, with the Scottish Highlands to the north sliding past the Lowlands to the south. But, as things stand at present, we cannot even say with confidence whether the Highlands slid into position from west to east (right lateral) or from east to west (left lateral). The most convincing argument, however, seems to be that the shift of the Highlands into position occurred in a right lateral sense. A comparison of the various ideas about the Great Glen which have abounded for the past three decades shows the kind of fascinating detective work for which there is still plenty of scope in modern geophysics.

Although the established view was that the fault line of the Great Glen, which runs across Scotland in a south-west–north-east direction, must have been formed by vertical movements of the Earth's crust which had cracked at a line of weakness, in 1946 the geologist W.Q. Kennedy made the then almost heretical suggestion that horizontal movements might in fact have been the key process. By itself, the evidence from Scotland was not enough to overturn the geological establishment and gain acceptance for the idea of continental drift, but once the new concept of plate tectonics had made the drifters reputable the Great Glen became an obvious case for treatment and a fine example of tectonic forces at work – even though no one was sure which way they had been working!

Ironically, Kennedy's suggestions were based on geological evidence which now seems to be inadmissable. He noticed that certain characteristic granite outcrops in the rocks north and south of the Great Glen could be related to one another if they had once formed one outcrop and been split by movement along the fault by some 75 kilometres, with the Highlands sliding left laterally to the south-west, about 65 million years ago. With the acceptance of the possibility of continental drift, these ideas became respectable and held sway until the early 1970s, when they were challenged by two members of the Institute of Geological Sciences in London. Dr M.S. Garson and Dr Jane Plant argued that although Kennedy had been correct in assessing the Great Glen as a transcurrent (sideways slipping) fault, he had got the direction wrong. On their model, the Highlands of Scotland have slipped into position in the opposite direction, moving to the right as viewed from England.

This new interpretation came about through the great increase in the amount of geological data available concerning the region of Scotland and the North Sea – some of it, of course, obtained as a direct result of the search for oil in that part of the world. The Foyers and Strontian granites picked out by Kennedy (see Figure 10) no longer seemed such a good guide to the slip, and probably never formed part of the same complex at all. Instead, Garson and Plant gathered and assessed an impressive weight of evidence about other rocks on land, and from surveys in the North Sea to the east and the Atlantic west of Ireland. These surveys show extensions of the fault line in both directions, with clear evidence that the rocks to the north have moved in an easterly direction along the fault.

Magnetic data, from one of the new areas of geophysical study, have been particularly valuable in this kind of

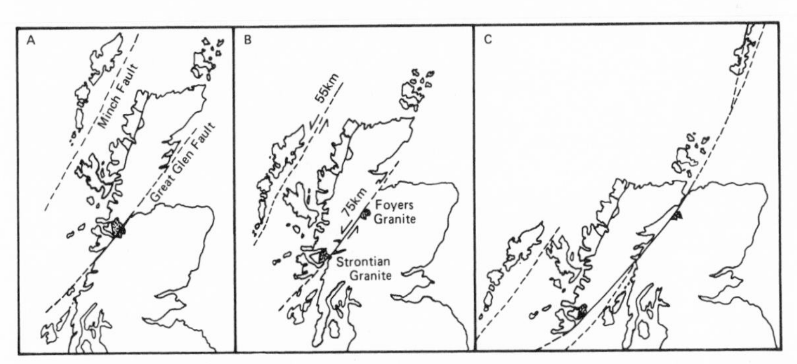

Fig.10 Two possible reconstructions of Scotland.
A Before the shift postulated by Kennedy.
B Present-day situation with Foyers and Strontian granites separated.
C Position before the shift postulated by Garson and Plant.
(Reproduced by permission of Dr Jane Plant and *Nature*)

research, and some twenty-five years after Kennedy's original proposal about the nature of the Great Glen Fault Dr D. Flinn, of the University of Liverpool, pointed out that the magnetic jigsaw puzzle of the rocks around the fault where it runs through the sea floor could be reconstructed satisfactorily only if the rocks in that part of the world had slid right laterally – in the opposite way to the slip postulated by Kennedy for the Great Glen itself. But Dr Flinn did not go so far as to throw out Kennedy's suggestions, and for a couple of years there was a confusing situation in which geologists seemed to accept that some parts of the fault had been involved in movements of sixty kilometres or more in a right lateral sense, while other parts (including Loch Ness) had been involved in even bigger movements in the opposite direction.

Of course, this couldn't last, and the reconstruction by Garson and Plant not only sorted out the confusion but also showed that the Great Glen Fault system can be traced not just out into the neighbouring sea around Scotland but even into continental Europe and North America. This modern picture sees the crucial event in the history of the Great Glen as a major movement of the Highlands of Scotland – a fragment of continental crust as large as England – by about 100 kilometres towards the north-east some 400 million years ago. At that time, the Atlantic Ocean was just beginning to form, and the pieces of continental crust we now know as Canada and Greenland were Scotland's near neighbours . . . So what evidence is there of related geological features in these parts of the world?

There are many fault systems in the Canadian Appalachian Mountains and in Newfoundland; one of these, the Cabot Fault complex, could once have formed one giant complex with the Great Glen and the Leannan Fault in Ireland, before the Atlantic opened. To the north-east, the line of the Great Glen Fault can also be extended, and may tie in with the sideways fault systems found in Spitzbergen. Although the movements along the Scandinavian faults are not yet understood, the evidence from the Cabot system again suggests a north-easterly slip, and overall there is impressive evidence that before the formation of the Atlantic

Fig.11 The Great Glen from Fort William looking towards Loch Ness. (Reproduced by permission of Aerofilms Ltd)

Ocean a fault system extending across 3,200 kilometres – as large as the present-day San Andreas system – extended from the primeval Appalachians through what is now Scotland to Scandinavia. The activity of the fault must have been related to the opening of the Atlantic (whereas the activity of the San Andreas is an early sign that the Pacific Ocean is closing) and the Highlands of Scotland are almost certainly a splinter of continent left behind by Greenland and Canada as North America separated from Eurasia.

Although the Great Glen is an old fault, now quiet, so that we cannot see any traces of creep today, it does have one advantage over the San Andreas for the casual visitor. Because it is an old feature, the Great Glen is weathered and provides a home for many lochs; these lochs are now connected by the Caledonian Canal, which makes it possible to traverse the entire length of the fault, from coast to coast of Scotland, by boat. The canal was one of the great engineering feats of the nineteenth century, although it never really succeeded in fulfilling all the objectives which stimulated the effort to build it. These ranged from the Baltic trade interests, wanting a quick, safe passage between the east and west coasts, to a desire to support the west coast herring industry, to halt the depopulation of the western Highlands and to provide a strategic route of value during the Napoleonic wars. This last stimulus was the least successful of all, since although Thomas Telford was engaged to supervise the work in 1800 the canal only opened in 1822; soon enough in view of the difficulties of the task, but rather late to have any influence on the Napoleonic wars (see R.N. Millman, *The Making of the Scottish Landscape**). After repairs in the 1840s, the canal was used by small coasters carrying coal, farm produce and building materials and by fishing boats; but by the early twentieth century this narrow canal with many locks (covering a total lift of thirty-two metres) was obsolete and it has not been used extensively since, except during the two world wars when its strategic value at last came into its own.

However, the Great Glen was important to the Highlands long before the days of the Caledonian Canal. This is, of course, the most prominent fault line of the many which strike south-west–north-east across the Scottish Highlands and it has been opened out by both glaciation and rivers to provide a barrier across the country. The continuation of the Great Glen Fault can be seen in the Mull and along the line of cliffs of the Black Isles and Tain Peninsula to Tarbat Ness, and the strategic value of such a barrier to north–south routes has long been clear. At the end of the twelfth century William the Lion built castles at Edindour and Dunskaith in Ross-shire and at Urquhart and Inverlochy, to control the Great Glen and secure Argyll; these were part of the Celtic effort to extend their influence beyond the Grampians, in opposition to the Norse Lord of the Isles, which shows just how little of Scotland was under 'Scottish' control 800 years ago.

Routes in the Highlands had to follow the line of the glens because of the rugged terrain, but the Great Glen itself was never a great route before the eighteenth century because of the high passes at Glencoe and Rannoch Moor which barred

* Batsford, London, 1975.

the way to Glasgow from the south-west end. The Great Glen also marked a boundary for the spread of technological and industrial innovation, with the 'modern' agricultural techniques the exception rather than the rule north of the glen as recently as 200 years ago. All this began to change after the 1715 Jacobite rising, when General George Wade initiated a wave of road building to improve communications and, hopefully, to pacify the Highlands. These roads included links between the three forts along the Great Glen and from the forts to Dunkeld and Crieff in the south (this link is now part of the A9). The military roads were turned over to civil authority in the late eighteenth century, but even a dozen years ago many 'major' roads north and west of the Great Glen were still single track with passing places. R.N. Millman points out in *The Making of the Scottish Landscape* that, ironically, the improvement of roads in recent years has made transport easier and thus encouraged the depopulation of the Highlands, where many recently abandoned crofts can now be seen.

The Highlands have also suffered from other errors of judgment, most obvious to the visitor as he stands in the shadow of the foreboding ranks of conifers planted between the wars. Although the Forestry Commission today is far more enlightened and has learnt from the mistakes of its predecessors, these awful reminders of misguided policy will be present for some years yet. In farming as well past errors are now slowly being rectified, and cattle can be seen grazing in the glens again, including the Great Glen, in experimental areas. Cattle graze less selectively than sheep and provide better manuring, bringing back the lush grass and ecological stability that were lost when sheep were introduced in the late eighteenth century.

All this is a far cry from the San Andreas Fault. However, one feature of the Great Glen provides a good example of how parts of the San Andreas might develop if the right lateral movement stopped in the near future and it was left to erosion to continue changing the shape of the rocks in the region. The scoured-out valley of the Great Glen today is a young feature in geological terms, being very late Tertiary (but pre-glacial) in origin; before then, Highland rivers flowed across the line of the fault, and their traces can be seen in the form of high level cols. The Gour, for example, once flowed by the pass at Glencoe and eventually into the Tummel; the Eil flowed into Glen Nevis and the Arkaig into the Spean Valley. But the line of the fault encouraged the development of linking flows joining the rivers and draining them along the fault (in geological jargon, this is known as a 'piratical' system; the extreme erosion which is so characteristic of the Great Glen is called 'beheading', which obviously follows naturally from piracy). Such a piratical system would inevitably develop in the San Andreas, if only the plates would stop moving, especially where the lines of displaced streams already produce right-angle bends with flows along the fault line for part of their path. With a piratical system established, the fractured rocks of the fault line provide a situation almost tailor-made for erosion, and in the Great Glen the flows have scoured deeply to produce the lochs and glens we see today. One side-effect of this is that rivers on the southern side of the Great Glen now flow 'backwards', into the glen, having reversed their direction (at least in the mountains bordering the glen) as the rift was eroded below the level of the old river. The depth reached by this erosion is impressive; Loch Lochy is 531 feet (163 metres) deep and Loch Ness 754 feet (232 metres), with its deepest point 700 feet (215 metres) below sea level – certainly an ideal home for a trapped aquatic monster, whether or not one actually lives there.

In both of our case histories, the San Andreas and the Great Glen, we have seen how the broad picture of global tectonics makes it possible to interpret the details of regional activity. Of course, the development of an understanding of the details also helps to improve the broad picture, as in the way the pieces of evidence accumulate to explain the bend in the chain of the Hawaiian Islands. But it is still very much the case that, as in the search for new oil reserves, it is the broad overview that provides the key insight; and, as far as man is concerned, the broad picture of our changing Earth may be important not just for our present global society but also for an understanding of the evolution of man and his civilization in the first place.

Chapter Eight
Continental Drift, Evolution and Ice Ages

The emergence of man's civilization has taken place since the end of the most recent ice age some 15,000 years ago and by and large at fairly high latitudes – north of the tropic of Cancer. It is widely accepted that the earlier spread of ice played a large part in allowing man to become the dominant species on Earth, by killing off less adaptable creatures and producing conditions in which man's unique adaptability, ingenuity and cunning were the best characteristics for survival in a situation of harsh and changing climate. If this idea is correct, our existence today may owe something to the structure of the Galaxy (as discussed in Chapter One), but it certainly also owes a lot to the processes of plate tectonics. The continents today, and throughout recent geological history, are concentrated in the northern hemisphere in just the right way to encourage the occurrence of ice ages, once some trigger – such as the astronomical effect – produces a slight cooling (see Chapter One).

Because the Arctic sea is almost landlocked, warm water cannot penetrate it easily and sea ice can spread. The surrounding land becomes covered in a thin layer of snow during the winter, and this light coloured covering reflects away a lot of the Sun's heat. If conditions changed just a little bit, it is easy to imagine that much of the snow might persist through the summer at high latitudes and that, by reflecting heat away, this would encourage more cooling and growing snow cover in a feedback leading to a new ice age. This is an oversimplification, but one which does seem to have more than a grain of truth. Detailed calculations of the change in albedo (reflectivity) of the Earth for different distributions of the continents show that in the extreme case, with all the land concentrated in a belt around the equator, the temperature might be as much as 12°C warmer than the average today, with land and ice caps concentrated near both poles and producing much more reflection. Among other changes wrought by continental drift, it is just possible that the death of the dinosaurs really was caused by the slow processes of plate tectonics changing the face of the Earth enough to change the climate beyond the limits to which they had become adapted during 200 million years of warmth and plenty. Such ideas are still a little speculative, but quite plausible and certainly entertaining. Even more speculative, but just as entertaining, are a couple of suggestions about how evolution may have been affected by another geophysical process, the repeated reversing of the Earth's magnetic field.

The first possibility is that this process had a very direct effect on evolution. Today, the Earth's magnetic field plays an important part in protecting us from the cosmic radiation of charged particles which floods through space, but during the stage of a magnetic reversal, when the field has died away to nothing, this protective shield is removed. An increase in background radiation reaching the surface of the Earth might be enough to prove lethal to some species, although this seems unlikely, or it might just be enough to increase the rate of mutations by damaging the genetic material of many species. An increased genetic mutation rate leads to an increase in the rate of biological evolution, and this might, so the argument runs, explain why the population of the Earth should suddenly show a phase of rapid development after tens of millions of years of stability. This idea looks a little implausible, however, in the face of measurements which show that even today the magnetic field only screens out about fourteen per cent of the background radiation, and that in any case cosmic rays and radiation from active material on Earth between them only account for about one in twenty human genetic mutations.

Still, there is some evidence from the fossil record that some species did disappear at the times of geomagnetic reversals, and a slightly more complicated (but rather more plausible) theory might well explain the link. This brings us back to climatic change, suggesting quite simply that when the Earth has no magnetic field, in the middle of a reversal, conditions are somehow right for an ice age to be triggered. The trick, of course, is to explain the 'somehow', and two suggestions have been offered. A temporary disappearance of the Earth's magnetism could be accompanied by changes in the ionization of the upper atmosphere, which could affect the vital ozone layer (of which more in Chapter Nine) and the circulation of the atmosphere, producing a sudden climatic change. But a rather better idea suggests that reversals in the Earth's magnetic field are accompanied by outbursts of volcanic activity, and that the dust from the volcanoes acts as a sunshield so that the Earth cools off suddenly at such times.

The vagueness of all these ideas shows how difficult it still is to relate long-term geophysical changes to the history of man and other life on our planet. It is quite likely that most of the ideas of this kind which are tossed around have an element of truth in them, but that no one process alone is sufficient to overturn the natural balance. Magnetic reversals, for example, have occurred twenty times in the past four million years, and many more times before that, and there is no evidence that *every* reversal is accompanied by changes in the fossil record or by climatic changes. It is probably when several factors act together that we get a significant change in population or climate – but it must be very rare that these many factors acting together produce a change as drastic as the end of the dinosaurs. However, even accepting that our picture of events earlier than the 1½ per cent of Earth history we call the Cainozoic must always be rather blurred, some of these general ideas can help to bring into focus the events of the past 70 million years or so which are of such importance from the human point of view. As a general example, the link between volcanic activity and climate is worth a detailed look; and there is also a wealth of detailed evidence of how recent continental drift has affected the evolution of specific species.

Volcanic ash is found in many sedimentary layers in different parts of the world and the relationship between widespread outbursts of volcanism and widespread cooling is quite marked, both when we look at the very recent past and at the past couple of million years. New Zealander Dr J. M. Bray has made a particular study of this relationship over the past 40,000 years, the limit in time over which the accurate radiocarbon (Carbon-14) dating technique can be used. He reported his findings late in 1974, showing that the dates of samples of volcanic debris show a significant tendency for eruptions to occur at roughly the same time in widely separated parts of the world. In particular, almost synchronous eruptions have occurred in Japan, southern South America and New Zealand on several occasions during the past 40 millennia. There have been eight major waves of volcanic activity in South America during that time, and all eight of them coincided with outbursts in Japan, while six of the eight waves also coincided with volcanic outbursts in New Zealand. Furthermore, the outbursts – and several lesser events in each of these particularly active regions – can also be associated with the spread of glaciation.

Even without the climatic link these discoveries are of interest in tectonic terms. All of the three regions mentioned by Dr Bray are on the boundary of the Pacific plate, and it looks very much as if stresses which build up in the Earth's crust lead to simultaneous fractures and volcanism, and presumably related earthquake activity, around the margins of that plate.

Going back 41,000 years, to the limit of reliable Carbon-14 dating, Dr Bray finds that large-scale eruptions took place just before a major advance of the polar ice. More recently, in the past 17,200 years there have been eighteen phases of volcanic activity and related glacial advance, and in all but four of these cases the delay between the volcanic outburst and the spread of ice is between 100 and 300 years. It seems likely that these lags provide a clue to the climatic processes which encourage ice to build up once the triggering effect of volcanic activity has occurred – but discussion of the details of the complex processes of climatic change are beyond the scope of this book (see bibliography).

Just a couple of months after Dr Bray's analysis of recent volcanic activity appeared in *Nature*, a study of global volcanism over the past 20 million years by two U.S. oceanographers, Professor James Kennett and Mr Robert Thunell, appeared in the journal *Science*. Their investigation shows a broader sweep than the detailed changes investigated by Dr Bray, and provides another insight into the cause of the most recent glaciation. You may wonder how oceanographers get in on this game; the answer is simply that cores drilled from the ocean bed provide some of the best evidence of the general background of volcanic activity, since dust which falls on the oceans is well mixed and well away from the local influence of particular volcanoes on particular continents. The cores which were studied by the U.S. team came from a worldwide survey, the Deep Sea Drilling Project (D.S.D.P.), and covered 320 sites from every ocean except the Arctic. They restricted their survey to the top layers of sediments, corresponding to the past 20 million years, because these are the layers in which the sediments can be best identified and dated. The changing level of volcanic activity revealed by measuring the distribution of the volcanic dust in these sediments is shown in Figure 12; the dramatic increase in volcanism during the past two million years ties in very well with the occurrence of the most recent ice age. Although the details from cores drilled at different sites around the world vary, they all show this recent increase in volcanic activity.

Some evidence from the land supports the idea that the past couple of million years have been unusually active in volcanic terms, and there seems little reason to argue with the conclusion reached by this U.S. team that 'increased Quaternary volcanism coincides . . . with that episode of the Cenozoic marked by major and rapidly fluctuating climatic changes.' However the D.S.D.P. survey is too coarse to pin down specific relations between volcanic outbursts and increasing ice cover, and although work like that of Dr Bray shows that increased volcanism can lead to increased ice cover it is also true that the ice cover sometimes develops without volcanic outbursts, and that sometimes volcanic outbursts occur with no apparent increase in ice cover

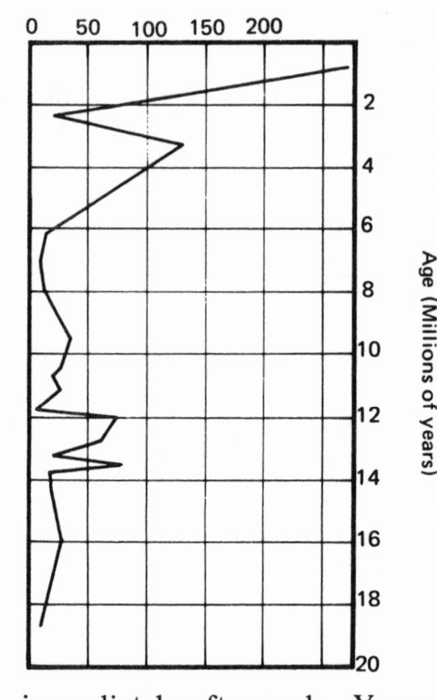

Fig.12 Variations in volcanic ash deposited over the past twenty million years. (After J.P.Kennett and R.C. Thunell, *Science*, vol. 187, 1975, pp.497–503.)

immediately afterwards. You can even turn the whole model on its head, and argue that during the past two million years the repeated loading and unloading of the Earth's crust by the weight of ice as glaciers have advanced and retreated has triggered volcanic activity and produced the patterns of ash layers found in the deep sea cores! And if you want to be really awkward, why not argue that both effects can occur, so that once either volcanic dust or ice cover increases there is a strong feedback by which the spreading ice triggers more volcanism which encourages further cooling? Whatever the truth of the matter, one thing is clear – it is impossible to solve the problems of Earth sciences which are of greatest interest to man by dealing with them in rigid compartments. When we are learning about plate tectonics, for example, it is convenient to ignore most of the interactions between the solid Earth and the atmosphere and oceans, as we have throughout most of this book. But when it comes down to applying that knowledge to practical problems, we need to take account of many different branches of the Earth sciences, and their interactions, to get a clear picture of what is going on.

I remain convinced that while volcanic dust certainly plays a large part in producing climatic change, it can only produce a great effect when conditions are just right. It has been known for a long time, for instance, that the changing tilt of the Earth – the tilt which is responsible for the differences between the seasons – varies so that sometimes there is very little difference between summer and winter while at other times there are extreme differences. With the present distribution of the continents, cool summers in the northern hemisphere encourage the spread of ice because snow which has fallen on the land does not melt so easily; and such cool northern summers always coincide with severe winters in the south, which encourages the spread of sea ice, also moving the climate towards ice age conditions. When you look at the whole history of the Earth, ice ages are not really very common, and it is clear that the past two million years or so have been rather exceptional. So it is not too extravagant to suggest that we have had periods of greatest ice cover when the continents are in the right place, when the tilt of the Earth is making a contribution and when the Solar System is in the right part of the Galaxy; and *then* we get a period of increased volcanism. Any two of the factors working together might do the trick, but I doubt very much that it is possible to explain the changing climate of our planet by variations in one process alone.

After this brief look at the complexities of the climatic problem it is time to return to the more simple situation of the direct effect of the breakup and drift of continents on the development of species which have become separated from their brothers, sisters and cousins by new, ever-widening oceans. One of the ironies of the way in which plate tectonic theory has helped the study of evolution is that Alfred Wegener himself originally put forward his ideas on continental drift because he learned of the similarity of fossils found in Brazil and Africa. This was – and is – a more persuasive argument than the easily seen 'fit' between the coastlines on either side of the South Atlantic, but even in the early twentieth century the evidence was still being explained away in terms of land 'bridges' which once existed between the continents but have now sunk beneath the oceans. Of course, this concept cannot stand up at all to the modern discovery that the thickness of crust beneath the oceans is much less than the thickness of continental crust, quite apart from all the other evidence in favour of drift that we have now accumulated. Now, in a happy example of scientific feed-back between different specialities, fossil evidence provides important clues to date the breakup of continents, and the processes of continental drift provide the palaeontologists with a new insight into the diversification of species around the world.

The evolution and distribution of land animals, in particular, is affected by two principal factors: the climate and the existence of water barriers, especially the oceans. Palaeontologists recognize three categories of dispersal route by which animals can spread from a local region, and although any such neat classification can only be approximate this provides a useful guide to the sort of hazards moving populations might encounter. The first category, 'corridors', allows animals to migrate freely in opposite directions, and the link between North and South America is an example. The second category, 'filter bridges', allows movement of some animals, but bars others. The link between North America and Asia which existed during the Pleistocene is an example; only animals which could survive the cold climate of the northern regions were able to make the journey across what is now the Bering Strait. The third category is the pot-luck possibility, 'sweepstake routes'. This covers the few creatures that manage to cross water barriers by chance on floating logs and so on – known as 'rafting'. Since only a few creatures can make such journeys, many islands have accumulated peculiar and unbalanced faunal populations; in islands cut off from the mainstream of evolution animals may still exist even though they have become extinct everywhere else, because the creatures that have displaced them from their ecological niche in the outside world have never reached their island havens. The biggest sample of this kind is provided by the many unique animals of Australia, where a very large land mass has been cut off from the outside world.

Continental drift has two major effects on populations, in genetic terms. When a continent splits apart, uniform populations are divided, and the two halves begin to follow their own evolutionary paths – they diversify. But when continents collide, the faunas of the two regions begin to interact and homogenize, with some groups dying out in the face of new competition. The classic example of convergence is not, in fact, the result of continental drift, but it is so neat that it is worth mentioning anyway.

For a long period, there was no connection between North and South America, even though the two continents occupied much the same relative positions as they do today. It was only at about the end of the Pliocene that the Isthmus of Panama formed, allowing migration in both directions (a 'corridor') and by that time there were, according to the best estimates, twenty-nine families of mammals living in the south and twenty-seven quite different families in the north. During the past three million years, the two populations have been able to interact and converge. Many northern mammals have moved south, while the armadillo has spread his range to the north, and today the two continents have twenty-two families of mammals in common.

At the same time as the Isthmus of Panama linked the two continents, it divided the Atlantic Ocean from the Pacific. As a result, marine populations in the shallow waters on either side of the isthmus began to diverge, so that many molluscs from the Caribbean, for example, are now noticeably different from their Pacific cousins, although clearly still related. This complementary pattern of divergence among marine creatures and convergence among land populations when continents are joined by new bridges (with the reverse happening when such bridges are broken) provides a valuable means of double-checking the fossil record and dating the tectonic events. But as yet it is far from easy to pin down precise dates from the evolutionary changes, since we simply do not know very accurately how long it takes for evolutionary processes to produce an observed degree of diversification in divided populations. Indeed, in parts of the world where the breakup has been dated very accurately by geophysical means palaeontologists are now trying to establish a more accurate evolutionary clock so that they can apply it reliably to date the breakup in places where the geophysical measurements are less certain. At least one creature, however, has been doing his best to overcome the effects of sea-floor spreading for 70 million years, and until recently managed to confound biologists in the process. The odyssey of the green turtle is rather a special case of the interaction between continental drift and evolution and also highlights some aspects of the modern theory of how the South Atlantic has evolved; it is also a fascinating story in its own right, so it is well worth telling in detail.

One family of the green turtle, known as *Chelonia mydas*, lives on the coast of Brazil but breeds and nests on Ascension Island, 2,000 kilometres away in the central region of the equatorial Atlantic Ocean. How and why these turtles should carry out the remarkable feat of endurance and navigation involved in this migration has long been a puzzle; but the puzzle may have been resolved in terms of sea-floor spreading by Dr Archie Carr of the University of Florida and Dr Patrick Coleman of the University of Western Australia, who presented their theory in a remarkable article in *Nature*, published in the Spring of 1974.

There is no doubt that the fundamental reason for turtles' preference for offshore islands as breeding sites is that these havens are by and large free from the egg-eating predators common on the mainland, where the turtles spend most of their lives. But a trek of 2,000 kilometres is a little more than the usual trip undertaken by other groups of turtles to offshore islands – indeed, it takes *Chelonia mydas* right out to the edge of the mid-Atlantic spreading ridge. And that, it seems, is the vital clue to the origin of their odyssey. According to the latest ideas of sea-floor spreading, the opening of the equatorial Atlantic and the separation of South America and Africa took place in a series of steps. First, about 110 million years ago (or perhaps even earlier) rift valleys formed; then, up to about 80 or 90 million years ago the valleys were flooded in phases from the widening oceans to the north and south; finally, about 80 million years ago, and certainly by 70 million years ago, the two oceans were joined and their spreading ridge systems linked in one continuous system, although with the equatorial region of the ocean still very narrow, hardly more than a channel, with a string of offshore islands along the new continental coastline.

Palaeontologists know, from fossil evidence, that there were turtles living in this oceanic channel, and although it cannot be proved that some of these turtles were the direct ancestors of *Chelonia mydas* it seems quite reasonable that they should have been. However, Ascension Island itself is only about seven million years old, so 70 million years ago these ancestors of the present turtles could hardly have been breeding there. But Ascension is just the latest in a series of volcanic islands produced by the activity of the spreading ridge, and its immediate predecessor lies below the surface of the sea fifteen kilometres away to the west. What happens, according to the tectonic ideas outlined in this book, is that each volcanic island is carried away down the side of the mid-Atlantic ridge into deep water, sinking below the surface as it goes, on the conveyor belt of spreading oceanic crust. Then, a new island emerges close to the ridge as volcanic activity continues – perhaps over a hot spot. The situation is reminiscent of that in the Hawaiian chain, although there the islands are bigger and, created in the deep ocean, remain above water as they drift away. Seventy million years ago, the volcanic islands of the mid-Atlantic ridge would only have been about 300 kilometres away from the coast of South America, assuming that spreading has been going on continuously since the Atlantic opened at a rate of about two centimetres a year. This would be a reasonable journey for any turtles which overshot the genuine coastal islands of the continent, and they could have bred there happily, with no realization in their turtle brains of the trouble they were storing up for their descendants.

As yet, there is no definite proof that there is a continuous chain of submerged volcanic islands – seamounts – from Ascension to the coast of Brazil, but both ends of such a possible chain can be identified on the ocean floor and there seems little reason to doubt that the middle of the chain is there as well. Since the spreading is roughly east–west, and the turtles have to swim almost along constant latitude in their migration today, the only change in the migration made necessary by continental drift is a gradual lengthening as successive volcanic islands have sunk beneath the waves and been replaced by new ones. But even at two centimetres a year, that lengthening adds up in 70 million years to the impressive journey the turtles make today.

Indeed, the migration is made especially simple because of the natural change in the position of the rising Sun over the period of the journey, which takes about eight weeks. For turtles leaving the coast of northern Brazil in December, a path straight into the rising Sun takes them initially east-south-east. Resting at night, the turtles have only to set off again each morning into the Sun for their path to curve gently northwards as the Sun rises a little further north each day. After a couple of months, the travellers end up 'downstream' of Ascension Island in the prevailing Equatorial Current, and can home in on it through smell or by identifying some chemical trace in the water flowing past the island. Over millions of years, it may be that the next generation of 'Ascension' islands emerges before the old island has slipped away beneath the waves, giving the turtles a choice of breeding sites for a time; they would certainly be in a great deal of trouble if the time ever comes when the replacement island does not emerge before its predecessor sinks away, and it is difficult to see how they will manage the journey as the Atlantic continues to widen for tens of millions of years more. It may be that we are witnesses of an extremely rare, possibly unique, example of adaptation of a species to the problem posed by sea-floor spreading. If the present-day family of *Chelonia mydas* really has inherited its pattern of behaviour over more than 50 million years, it has certainly had ample time for natural selection to have evolved the present form, with enormous shoulder muscles and large fat deposits necessary for it to make the regular migration. In providing us with an example of a species in which the seaward migratory drive has directly caused the evolution into the form we see today, these green turtles warn against too glib an assessment of the species of our planet by their obvious external features alone. This may be an extreme case, but all the animals of our planet, including man, have been moulded by the changing geography and the changing climate (often with the two changes going hand in hand). We are very much the creatures of one small planet, our home in space – Spaceship Earth.

Chapter Nine
Spaceship Earth

We have now seen how the interior of the Earth is structured and how stirrings in the outer layers move the plates of the crust about in an ever-changing pattern of continental drift. But if we look at the Earth as a planet in space there is still one important outer shell even beyond the crust of the solid surface – the biosphere in which we live. Taking the analogy of 'Spaceship Earth' to its logical conclusion, the hull of our spaceship is not the solid surface but the magnetosphere which forms the upper limit of the Earth's atmosphere and marks the limit of the influence of our planet's magnetic field. Important though the oceans are, man is today very much a land-based animal, and the oceans are still only poorly understood, so I shall take up the description of the outer layers of Spaceship Earth – the living quarters – from the base of the atmosphere.

The layered structure of the atmosphere of the Earth is best seen by considering the variation of its temperature with altitude (Figure 13). The source of heat which keeps the atmosphere warm is primarily the Sun, although a little heat does, of course, leak out from the warm interior. Some solar radiation is reflected into space at the top of the atmosphere, ultraviolet and infrared wavelengths are absorbed in the upper regions, but by far the bulk of this incident energy penetrates to the lower layers and reaches the ground. The warm ground then provides heat for the atmosphere immediately above it, a little of this heat by conduction but most of it by re-radiation at infrared wavelengths. Because these wavelengths are strongly absorbed by water vapour and carbon dioxide the lower atmosphere warms up, and again

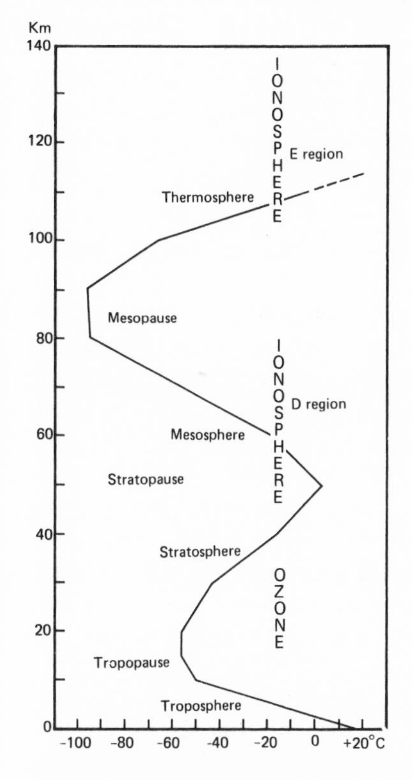

Fig.13 Variation of atmospheric temperature with height. (Based on Fig. 3.2 of *Everyman's Astronomy*, edited by R.H. Stoy [J.M. Dent, London, 1974])

energy is re-radiated, some going back down to the ground for another trip around the cycle, some going upwards to warm higher regions of the atmosphere. The net result of this bouncing around of energy is that the surface of the Earth is rather warmer than it would be if it received the same amount of direct radiation from the Sun but there was no blanket of atmosphere above; this is the 'greenhouse effect', often referred to but seldom explained. The heat also produces convection in the atmosphere, at least in the lower regions, and this plays a key part in the circulation patterns of weather and climate. But to a large extent this convection is confined within a well-defined layer near the Earth's surface, the troposphere, inhibited by the presence of ozone in the stratosphere above.

Through the troposphere, the layer in which weather occurs, the temperature of the atmosphere falls by about

6°C for every kilometre increase in height. This decrease slows down near an altitude of 10 kilometres, stops near 15 kilometres and from 20 kilometres to 50 kilometres temperatures increase from a minimum of about –60°C to a maximum of 0°C at the top of the stratosphere (the stratopause). This warming layer – the stratosphere – corresponds closely to the layer of ozone concentration, and it is the presence of ozone which is responsible for the warming.

Ozone is a blue gas with a distinctive pungent odour familiar to anyone who has worked with electric sparks, sniffed round the back of a colour TV set or used ultraviolet radiation for sterilizing in the 'germicidal' band of wavelengths below 2,800 Ångströms. Ironically, in view of its associations with health and vitality and fresh sea breezes – and its importance as a shielding layer in the atmosphere – ozone is highly poisonous, toxic even at a concentration of one part per million in air. The gas is highly oxidizing, which is no surprise since each molecule of ozone contains three oxygen atoms, compared with the two of ordinary molecular oxygen. Ozone absorbs radiation in the range 2,800 Å and shorter wavelengths, and it was first identified in the atmosphere because of this strong absorption of solar radiation; that absorption is what warms the stratosphere, but the details of the process are not quite as straightforward as they might seem at first sight.

First of all, molecular oxygen is dissociated into component atoms by absorbing ultraviolet radiation from the Sun. The amount of energy absorbed, and the number of molecules dissociated, depends both on the intensity of radiation available (which increases with altitude) and the number of molecules around (which decreases with altitude), so the effect is strongest where there is a balance between the two factors. When oxygen atoms are produced, free atoms can combine with molecules that have not dissociated to produce ozone (O_3), as long as there are plenty of ordinary oxygen molecules around to combine with, which again restricts the range of altitudes over which the process takes place.. In addition, this combination seems to be more effective when there are other molecules around, which catalyse the reaction without being affected themselves, and it is the combination of all these factors which restricts the range of ozone production so that the greatest concentrations are found between altitudes of 20 and 30 kilometres. The ozone itself can and does absorb radiation, so that it too dissociates to produce one free oxygen atom and one diatomic molecule from each ozone molecule; all these processes are going on at the same time, producing an equilibrium in which the amount of ozone in the stratosphere is roughly constant even though more molecules are being added, and others dissociated, all the time. It is like a river, which is always present between the river banks even though the molecules of water which make up the river are always changing as it collects run-off from rainfall and flows to the sea.

In spite of these complexities, calculations of the equilibrium balance have been made, and they agree well with the observations of ozone concentration made by measuring its absorption (from rocket- and balloon-borne instruments) for altitudes above ten kilometres. But because the pattern of reactions is fairly complicated, the way in which ozone concentrations vary as conditions change is not always obvious, and this has caused a great deal of confusion amongst those people who are worried that man's polluting activities may disturb the natural balance, removing ozone and letting harmful radiation through to the surface of the Earth.

Even the time taken for equilibrium to be reached if the balance is disturbed depends strongly on the altitude, and although it only takes a few minutes to reach equilibrium at altitudes above fifty kilometres it can take several days in the region below thirty kilometres. Because of this, the lower region is never really in equilibrium, being disturbed by meteorological circulation; since this region is just the region of greatest ozone concentration, ground-based observers looking up through the atmosphere see a strong correlation between ozone absorption changes and changes in circulation and weather patterns. More obviously, perhaps, the change of intensity of solar input with latitude and with the changing seasons also affects the ozone concentration; less obviously, however, even though ozone is produced by sunlight there is no tendency for it to go away at night.

Indeed, the overall ozone concentration tends to *increase* slightly at night, and this rather surprising result emphasizes the dangers of drawing 'obvious' conclusions about the behaviour of the layer. The reason for this particular effect is that at altitudes above forty kilometres the strong solar ultraviolet radiation dissociates ozone efficiently, so that when the radiation is absent the chemical balance shifts towards ozone. At lower altitudes, there would be a tendency for ozone concentration to decrease in the absence of sunlight, but there the time taken to reach the new equilibrium is far longer than one night, so there is no chance for the effect to become noticeable. The overall effect, through a column of atmosphere, is that the densest ozone concentration hardly changes at night, while in the tenuous upper layers ozone concentration increases. There is no doubt of the importance of the layer of strong ozone concentration to man; life only emerged from the oceans on to the land after enough oxygen had been released to build up an ozone layer which screens out intense ultraviolet radiation, and it is quite likely that this is more than a coincidence. Although we can't be sure what effect some of man's pollutants – high-flying supersonic jets or long-lived gases from aerosol cans – may have on the ozone equilibrium, it certainly seems wise to avoid taking risks with our valuable sunshield.

Above the stratosphere is another layer (between altitudes of 50 and 80 kilometres) in which cooling dominates; this is the mesosphere, and the lowest atmospheric temperatures (about −100°C) are reached at the top of this layer, between altitudes of 80 and 90 kilometres. From there outwards, the temperature increases steadily, mainly through absorption of energy by oxygen molecules as they dissociate, but here the shortage of free molecules and the intensity of the radiation prevent the formation of ozone. Rather, the absorption goes even further, with atoms being ionized, losing one or more electrons and forming an atmosphere of charged particles (positive ions and negative electrons). In a sense, the top of the layer of increasing temperature (the thermosphere) is at the 'temperature' of interplanetary space, but at such low densities the concept of temperature is no longer useful. A better guide to the limits of the atmosphere is provided by the estimate that by an altitude of about 500 kilometres the atmosphere is so tenuous that collisions between its component molecules, atoms and ions have become too rare for it to be treated as a continuous gas, and above this limit, in the region known as the exosphere, material can leak away into space. However, the best guide is provided by considering the ionized regions of the atmosphere in terms of the interaction with the Earth's magnetic field, in the regions where for charged particles magnetism, rather than gravity, is the main force holding them in a localized region.

The ionosphere is the region of the Earth's atmosphere extending from the stratosphere to the exosphere. The presence of ionized particles in this region was first suspected when it was discovered that radio waves can be transmitted 'round the corner' of the spherical Earth; radio waves travel in straight lines, and this long-distance transmission over the Earth is possible only because the ionized layers reflect radio transmissions at wavelengths longer than about fifteen metres. Shorter waves penetrate without being reflected, and this is why V.H.F. and TV transmissions can only be received if the receiver is within the line of sight of the transmitter. By this definition, the mesosphere and thermosphere are subdivisions of the ionosphere; but it is more usual to divide the ionosphere into three layers, D, E and F, defined by the extent of ionization within them.

The D region, from 50 to 90 kilometres, is weakly ionized and plays little part in reflecting radio waves. The E region, between 90 and 160 kilometres, contains more strongly ionized molecules, and the F region above contains strongly ionized atoms. The E region is distinguished from the F region in practical radio terms because its weak ionization almost disappears at night, which is why it is possible to pick up a different range of distant A.M. stations at night. These two layers are also known as the Heaviside-Kennelly (E) and Appleton (F) layers, after their discoverers, and the F layer is sometimes further subdivided into F1 and F2 layers. More important to the Earth scientists than the subtleties of these subdivisions, however, is the fact that both E and F layers are strongly affected by changes in solar activity, including individual solar flares, the 27-day period of rotation of the Sun and the roughly 11-year 'sunspot cycle' of solar activity. This influence of solar activity deep into the Earth's atmosphere may play a part in producing the changes in weather and climate which have also been linked with solar changes.

Beyond the ionosphere, the magnetosphere is the outermost region of the Earth's atmosphere, above an altitude of 500 kilometres where the ionization is so complete that the particles form a plasma constrained by the Earth's magnetic field. The magnetosphere forms the absolute limit of the Earth – the hull of the spaceship on which we live. On the 'upstream' side of the solar wind of charged particles which blows outwards past our planet, the interaction of the Earth's magnetic field produces a shock wave at a distance of about fourteen Earth radii. Closer in, two zones of high particle density are centred above the equator at heights of 3,000 and 15,000 kilometres; these are the Van Allen Belts, discovered in the 1950s from experiments on board some of the first artificial satellites. In the 1970s, study of the magnetosphere received a great boost from observations made by the European satellite HEOS-2, launched in 1972, which explored the magnetosphere above the North Pole, discovering the polar cusp (see Figure 14); it also discovered and mapped the magnetosheath, a long tail or mantle of plasma stretching away 'downstream' in the solar wind. In shielding the Earth from the charged particles of the solar wind, the magnetic field deflects them into the Van Allen Belts, but some spill over into the polar regions of the upper atmosphere, near the cusps, where fast-moving electrons interact with atmosphere atoms to produce the coloured lights of the aurorae. Associated activity alters the ionosphere and affects radio transmission near the poles, and through some mechanism which is still not fully understood the circulation of the atmosphere is also affected by bursts of solar particles so that a few days after bright aurorae cyclonic disturbances (depressions) at high latitudes are intensified.

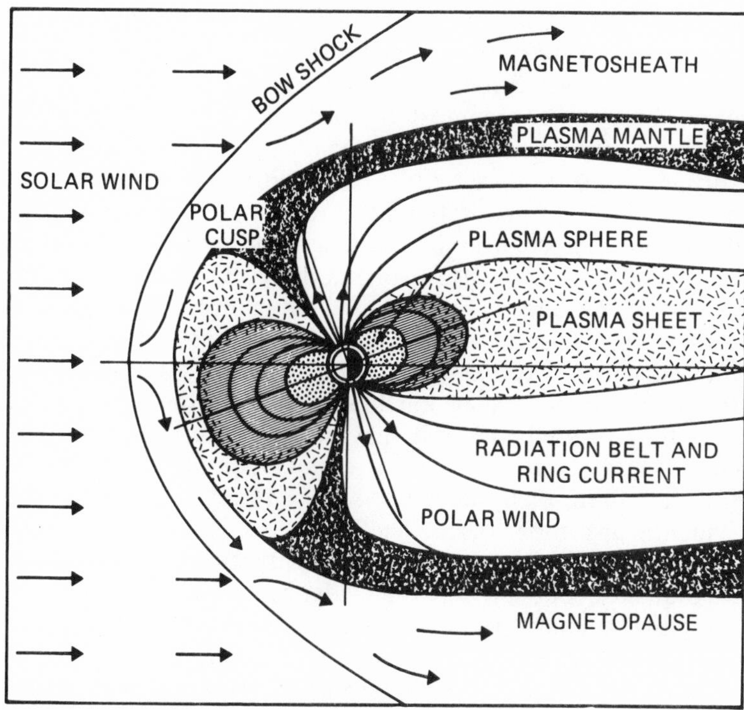

Fig.14 Noon/midnight cross-section through the magnetosphere showing the plasma mantle discovered by Europe's satellite HEOS-2.

The discovery of such interactions between the changing level of solar activity and the weather emphasizes the chief difference between our planet and a man-made spaceship. The power plant of Spaceship Earth is not within the hull, but outside: it is the Sun, to which we are connected by a gravitational leash. Without the Sun's heat, our spaceship would be uninhabitable, and since it now seems that the Earth responds to even quite modest solar hiccups it is clear that our more elaborate plans are at the mercy of the Sun's minor fluctuations and that these should be borne in mind when long-term plans are being made.

I have discussed the link between solar activity and climate in my book *Forecasts, Famines and Freezes*,* so here I will look instead at the direct link between solar events and

* Wildwood House, London, 1976 and Walker, New York, 1976.

seismic activity, which is of more immediate relevance to the changing face of the solid Earth on which we have focussed our attention throughout this book. This link has been discovered – almost against the better judgment of some of the discoverers – from separate evidence that changes in seismic activity are sometimes associated with changes in the spin of the Earth (changes in the length of day) and that changes in the length of day are sometimes associated with changes in solar activity. Combining the two pieces of evidence hints strongly at a link between solar activity and seismic activity, but the thought of suggesting that sunspots may cause earthquakes seemed so outrageous to many establishment Earth scientists in the early 1970s that they refused to admit that the evidence was more than coincidental. There is still a school of thought that argues this, but there really does not seem any reason to doubt the evidence any more, as we shall see. It might be outrageous to claim that all changes in solar activity cause significant earthquakes, but all that the evidence shows is that there is a slight tendency for an increase in seismic activity at certain phases of the sunspot cycle, and that the very largest flares of activity on the Sun can be correlated with tiny microseisms on Earth. This is not so outrageous as to offend the sensibilities of the seismologists, especially since the evidence of how changes in solar activity affect the circulation of the Earth's atmosphere provides an explanation of the physical link between solar activity and geophysical processes.

Since the early 1960s, with the advent of atomic clocks which provide sufficient accuracy for these studies, several groups of astronomers around the world have studied the variations in the length of day (L.O.D.) caused by the changing spin rate of the Earth. It is no surprise that the Earth is, on average, slowing down (with L.O.D. increasing); that effect has long been known and can easily be explained as a result of tidal action, especially the tides involving the Earth, Moon and Sun. But more subtle changes in the L.O.D. have now been found superimposed on this essentially steady slowing down, and these pose a variety of puzzles. One is the Chandler Wobble, in which the poles dance around in a stately fashion relative to the stars, affecting the L.O.D.

with a regular fourteen-month periodicity. The cause of this wobble is far from understood, although it may be associated with fluid motions in the Earth's core, and it is only predictable on the basis that it has been going on regularly for as long as it has been known, so there is no reason to suppose that it is going to change now. Earthquake activity also shows a periodic fluctuation (taking a global average) over about the same period, and the energies involved in the changing levels of seismic activity are about the same as those associated with the Chandler Wobble. So, naturally, some geophysicists argue that the seismic activity drives the wobble, while of course others argue that the wobble triggers earthquakes. It certainly seems reasonable that if the whole Earth is shaken then areas in which strain has built up in rocks could be triggered into seismic activity, and this idea is the recurring theme in investigating the way sunspots may, in fact, cause at least micro-earthquakes.

A more rapid variation in the L.O.D. occurs with the changing seasons, and there seems little doubt about the cause of this one, which lies in the large-scale movements of the atmosphere and changing seasonal wind patterns. This might seem a little surprising at first sight, since the atmosphere, as we have seen, is a very thin layer on the surface of the Earth. But because it is on the outside, even its relatively small mass can produce a measurable influence on the L.O.D. because its angular momentum – its leverage – is proportionally greater than that of the same mass at the core. The effect is similar to the way in which the spin of an ice skater can be changed by moving the arms around, but here we are talking about a variation of only twenty milliseconds or so over the year. So the atmosphere can affect the L.O.D., and this is the other link in the chain between sunspots and earthquakes.

The final link, between sunspots and the weather, is now firmly established through studies ranging from the specific effects of individual flares mentioned previously to changes in climatic factors and agricultural production over the eleven-year cycle and longer cycles of solar activity. So the discovery of a very small tendency for the level of seismic activity to be increased at times of sunspot maximum and minimum (the times when the level of solar activity is changing most rapidly from one state to another) is no longer a shock, even if it is not very useful in predicting specific earthquakes. The icing on the cake of this discovery of a close solar–terrestrial link comes from studies of a remarkable outburst of solar activity which occurred in 1972 – the biggest solar flare ever known, and large enough to produce a measurable effect on L.O.D. and on micro-earthquake activity all by itself.

Classic definitions of the standard of time, Universal Time or U.T., depend on measuring the rotation of the Earth relative to the fixed stars, and U.T. is still monitored continuously at observatories such as the Royal Greenwich Observatory at Herstmonceux in England, and the U.S. Naval Observatory in Washington. Both observatories have naval links, dating from the days when the navies of the world had to encourage research into accurate methods of timekeeping in order to improve navigation; today, measurements of U.T. are of vital interest to astronomers, who need to know how U.T. deviates from the new standard of Atomic Time (A.T.) in order to know just when to point their telescopes in a certain direction to see a certain celestial object. So continuous records of U.T., calibrated against A.T., are readily available to any scientists who contact these respected observatories. When the greatest solar storm ever known occurred in August 1972, I was already interested in the possibility of a link between such solar activity and earthquakes and was working on the problem with a colleague then doing research for NASA, Dr Stephen Plagemann. Waiting with some impatience for the end of the year, we then obtained a continuous record of the changing difference between A.T. and U.T. for the period May to October 1972, from the monitoring team at the U.S. Naval Observatory. This showed a more or less steady drift over the year, as A.T. and U.T. move steadily further out of step, with one clear peculiar feature – an abrupt jump just after the great solar storm.

The storm itself began on August 2nd and lasted for a few days; between August 7th and 8th an abrupt change in the spin rate of the Earth occurred, affecting the L.O.D. by more than ten milliseconds, the biggest change in one day over the entire period covered by the data. For the next few weeks, the spin of the Earth slowed rather less quickly than before the jump, but eventually returned to the same rate of steadily decreasing L.O.D. as before the solar flare activity (Figure 15). The short interval between the maximum of solar activity and the jolt of the Earth fits beautifully with the idea that particles and plasma from the Sun, funnelled on to polar regions by the Earth's magnetosphere, disturb the atmospheric circulation just enough to produce the observed effect on spin and L.O.D. When this suggestion was published, we received some criticism from the professional students of the changing length of day, who objected to our use of data from only one observatory and to the lack of 'smoothing' in the observations we used. Even now, many astronomers do not regard the evidence as conclusive. But the sort of effect we were looking for is just the kind of thing which standard smoothing techniques might obscure,

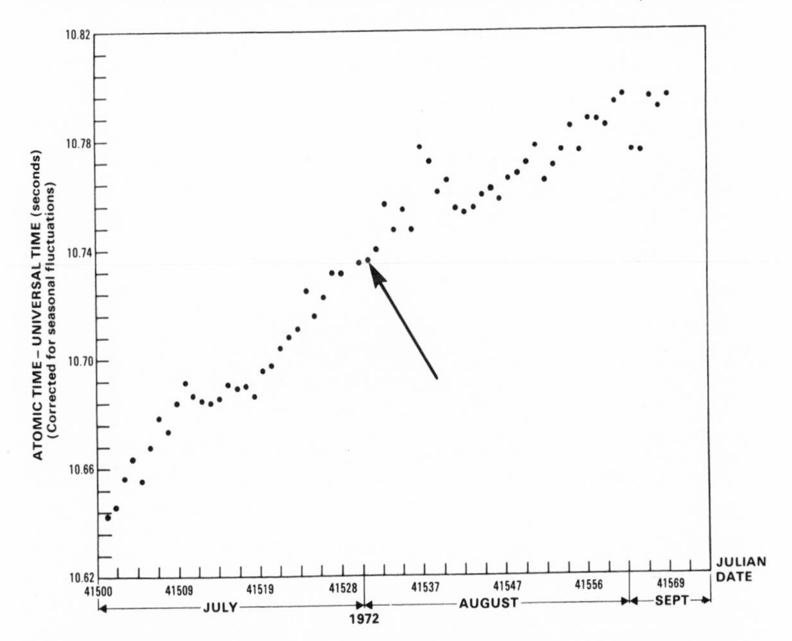

Fig.15 Variation in A.T.–U.T. during part of 1972. The arrow marks the time of commencement of the large burst of solar activity.

and although the criticism of our data from a single observatory is valid, it would be an extraordinary coincidence if exactly the effect predicted could be found in the data from that one observatory, in just the right place, by accident. All this discussion, however, seems now to have been settled by an earthquake study carried out by Dr Surendra Singh, of the University of Science of Malaysia, which was reported in 1975.

This study concentrated on the time variation of micro-earthquakes near Socorro, New Mexico, and covered the period of the great solar storm of August 1972. Dr Singh found a sharp increase in the number of micro-earthquakes occurring just after the storm, with a time lag of about fifteen days producing the peak earthquake activity on August 20th, during the period when the rate at which the Earth was slowing down seems to have been affected. Micro-earthquakes with energies up to 10^{12} ergs at average focal depths of about five kilometres occur roughly twice a day in this area; on August 19th, 1972, there were fourteen such shocks, and on August 20th there were sixteen. In line with standard scientific practice it would be good to have reliable statistical evidence in the form of similar studies of similar changes associated with other bursts of solar activity – but since the August 1972 event was the biggest ever known, and the only comparable flare activity occurred in 1959, it is difficult to see how that can be arranged. The cynics still have room to plead coincidence, but the overall picture from many related areas of study is convincing – solar activity affects the weather, the spin of the Earth and even, to a small extent, earthquakes. It is just possible, perhaps, that if really large bursts of solar activity could be predicted this might be of value in earthquake prediction; this possibility is investigated in *The Jupiter Effect.** However, in our look at Spaceship Earth as a whole, we should now move beyond our own magnetosphere to see how changes in the solar magnetic sector structure of the interplanetary medium may pinpoint the nature of this solar-terrestrial link.

* John Gribbin and Stephen Plagemann, *The Jupiter Effect* (Macmillan, London, 1974 and Walker, New York, 1974).

The best guide to this comes from the study of the solar influence on weather, which was pioneered in its modern form by Dr Walter Orr Roberts, of the University of Colorado and the National Center for Atmospheric Research in Boulder, Colorado, working in recent years with several colleagues. To put this work on a sound statistical basis a parameter known as the vorticity area index is used rather than any subjective assessment of good and bad weather; basically, when this index (V.A.I.) is high then there are more and/or deeper low pressure troughs. The variations in this index have now been tied not just to any phenomenon in the ionosphere or magnetosphere that could perhaps have a terrestrial origin, but to changes in the solar wind itself, and in particular to features known as solar magnetic sectors.

Observations from space probes since the mid-1960s have shown that the influence of the Sun's magnetic field extends through space with a well-defined structure. If the Sun did not rotate, you might imagine the magnetic field lines blown radially outwards by the solar wind of charged particles to form pie-shaped sectors intersecting the Earth's orbit; with the Sun's rotation added, the boundaries of the sectors would be twisted like the flat spring in a clock. This is an oversimplified picture, but it gives a rough idea of what is meant by the solar magnetic sector structure, and the space probes show that the magnetic field direction has opposite signs (either towards or away from the Sun) in adjacent sectors, with a very thin dividing boundary. Since the direction of the magnetic field plays a big part in controlling the flow of charged particles, the passage of a sector boundary across the Earth's magnetosphere is clearly an important event in terms of solar–terrestrial links. At most times, there are four sectors, each covering an angle of about 90° as viewed from the Sun, and rotating as the Sun rotates every twenty-seven days, to produce a passage of a sector boundary across the Earth every seven days or so. The passage of a sector boundary causes an increase in geomagnetic activity one or two days later, and also affects the V.A.I.

For measurements of the V.A.I. at the 500 millibar pressure level in the atmosphere, which provide a rough guide to cyclonic activity, studies in the northern hemisphere show a clear pattern of the average activity after the passage of a sector boundary, although of course this statistical picture does not necessarily apply to every particular case. The index reaches a minimum in the first day after the passage of a boundary, then increases by about ten per cent during the next two or three days; even initially sceptical meteorologists at the University of Toronto found when they tried to show that this link is not real that the statistics cannot be refuted, and from being opponents of Roberts's theory they have now accepted it and continue research into details of the mechanism by which it operates. Meanwhile, Roberts himself has taken things a step further by tying in changes in the V.A.I. to solar flare activity.

This pattern is a little more complicated, but seems just as real. By day one or two after a solar flare, the vorticity area index has increased by five to ten per cent above the background level; by day two or three the associated geomagnetic storm has started; by day three or four the V.A.I. decreases by five to ten per cent below the background (perhaps as much as twenty per cent below the peak reached just after the flare); and by day five or six things are back to normal. Again, this average statistical picture is subject to variation on individual occasions, and of course there are times when the sector structure effect and the effect of a solar flare combine. But although a great deal of work remains to be done to sort out a physical theory to explain the changes in detail, there is no longer any room to doubt the reality of the effect. This must be particularly gratifying for Dr Roberts, who has been arguing almost as a lone voice that there is such a link between solar activity and the weather for twenty years, since before we knew that the Earth had a magnetosphere or that the Sun had a magnetic sector structure stretching out past the Earth.

There are two ways for this work to develop. On Earth, studies from the southern hemisphere are urgently needed so that the growing band of researchers investigating these changes can get a better overall picture of what is happening in global terms. And from beyond the hull of Spaceship Earth there is an equally urgent need to find out just which solar flares are most likely to produce marked changes in the vorticity of the Earth's atmosphere at high latitudes. At the time of writing (October 1975), just this kind of evidence has been reported by a research group in Taiwan. They have looked particularly at the link between large sunspot and flare groups and geomagnetic disturbances, but as we have seen the geomagnetic effect is part of an overall pattern including weather changes and perhaps microseismic activity. It seems that there is a definite association between spot groups and the associated flares occurring between 6.7° and 19.9° East on the solar disc and the occurrence of large geomagnetic disturbances on Earth. Allowing for the 27-day rotation of the Sun, this means that the active regions most likely to be effective over this solar–terrestrial link are in regions about one day east of the solar meridian. It may be that these particular active regions are well situated with regard to the magnetic sector structure for the full force of their effects to reach the Earth; certainly it seems that the pieces of this particular jigsaw puzzle are now rapidly falling into place.

So even when we try to look at our planet as a whole, we find that it is still impossible to draw a sharp dividing line between Earth science and astronomy. The ties with the Sun are much too strong for that, and perhaps that is a good thing, for it reminds us that the Earth is, after all, just a planet, only one among nine planets orbiting our Sun, and who knows how many planets orbiting who knows how many other Suns out in space. Sitting on our own home planet it is sometimes difficult to sort out the wood from the trees when trying to picture the overall situation. But now that man's space probes have visited other planets in the Solar System, sending back a mass of detailed information about at least some of their properties, we can look at the overall situation in a new light. The new understanding of the Earth helps planetary scientists to interpret the data from other planets, and to build up a picture of their similarities and their differences. Equally, the new understanding of the planets of the Solar System in general has been a great help in developing the understanding of our own planet, the special case of greatest importance to man, even if it is of little significance on the cosmic scale of things. Even ten years ago it might have seemed a little strange to include a discussion of other planets in a book about geophysics; now, it seems both natural and essential to stretch the literal meaning of the word away from its roots to cover the physics not just of the Earth but of other planets as well.

Chapter Ten
'Geophysics' of other Planets

Observations from spacecraft in the 1970s have shown that the relatively small inner, or 'terrestrial', planets of our Solar System are very similar to one another in basic form, even though they appear superficially very different to a life-form that is used to the wind and waves of Earth. These four planets – Mercury, Venus, Earth and Mars – (and a fifth if you include our Moon) are all rocky, solid objects. Their differences are chiefly of degree, caused by differences in size and the different distances of their orbits from the Sun. These are the planets which should be covered by our extension of the term 'geophysics' beyond the confines of the Earth. The gas giant planets, out beyond Mars, are quite different objects, and tiny Pluto at the edge of the Solar System is probably only an escaped moon of no great significance. But even restricting 'geophysics' to the study of terrestrial planets might be too narrow a view to take; in one respect, the possession of a large magnetic field, the planet most like the Earth is the largest of all the planets in our Solar System, Jupiter! So although the inner planets have been the focus of attention lately, and must form the basis of this discussion, I shall also include Jupiter as a reminder that the exploration of the outer Solar System is only just beginning, and that in many ways even keeping within the orbit of Mars only gives a parochial view of the Universe. However, it makes sense to begin with the inner planets, and indeed with the innermost of them all – Mercury.

The remarkable pictures from the spacecraft Mariner 10, which flew past both Venus (of which more later) and Mercury, presented planetary geophysicists with an intriguing puzzle – although Mercury is like the Moon on the surface, its interior must be more like that of the Earth. The surface cratering of the planet is so similar to that of the Moon that it would take a real expert to say with confidence that the Mariner 10 photographs really did not come from the Moon, if he had no evidence to guide him. Like the Moon, Mercury seems to possess two types of characteristic terrain, densely cratered 'highland' regions and less cratered plains rather like the so-called lunar maria. Craters are also found on Mars, by radar observations through the clouds of Venus, and on Earth, even though our home craters have been much scoured away by erosion. For the astronomer, this is a key piece of evidence showing that all the inner planets were subjected to an intense bombardment of rocky debris as they formed early in the history of the Solar System. But to the geophysicist, an even more exciting discovery is the presence on Mercury of many shallow cliffs, fault structures known as lobate scarps, which seem to be just the kind of features that would be produced by compression as the planet has shrunk slightly in radius over geological time. This hints strongly at steady cooling of a large iron core, which ties in with observations of the planet's magnetic field.

Because Mariner 10 was injected into an orbit around the Sun which lasts just twice as long as the time it takes Mercury to orbit the Sun once, the spacecraft returned close to the planet every two Mercurian years. This gave the scientists operating the spacecraft experiments three opportunities for detailed studies before the spacecraft steering fuel ran out, and comparison of the results provides a good insight into the structure of the planet's magnetic field.

The first surprise was that Mercury possesses a weak magnetic field, which interacts with the solar wind in a small-scale version of the way the Earth's magnetic field behaves. The field is only about one per cent as strong as that of the Earth, but even that one per cent is more magnetic field than either Venus or Mars possesses, and its discovery was sufficiently intriguing for NASA scientists to use precious fuel to target the third, and last useful, pass of Mariner 10 to skim just 327 kilometres above the surface of the planet and as close as possible to its north pole. This confirmed the strength of the field and its dipolar nature, hinting strongly that it is being generated inside the planet by the same kind of dynamo process which generates the Earth's field. But the planet spins so slowly on its axis, three times in two of the planet's 'years', that geophysicists have been quite taken aback to find the dynamo process effective, even accepting the planet has a large iron core. Clearly, insights from other planets have much to tell us about fundamental processes also operating on Earth.

The iron core of Mercury is proportionally very big indeed. Although Mercury has a radius of only 2,440 kilometres, compared with the Earth's 6,370 kilometres, its iron core must make up 80 per cent of the planet to account for its measured density – and that makes the core, with a radius of 1,800 kilometres, bigger than the inner core of the Earth, and more than half as big as the whole core. This is undoubtedly because Mercury is so close to the Sun that even more of the lighter material from which the planets formed was blown away by the Sun's heat and the solar wind than in the case of the Earth. It is as if Mercury is the left-over heavy material from a protoplanet which would have been at least half as big as the Earth if it had condensed at our distance from the Sun.

If it is the magnetism and interior of Mercury that have the greatest interest in geophysical terms, it is the outer regions and atmospheres of Venus and Mars that make those planets a focus of attention. Venus, the next planet out from Mercury, is still the most difficult of the inner planets to describe, because observations have been hampered by the planet's layer of cloud. Several probes have dropped sensors to the surface, which revealed that the atmosphere is very hot and very dense; radar studies show that the surface is cratered; and en route to Mercury Mariner 10 sent back pictures of the cloud patterns of Venus and confirmed the absence of any significant magnetic field. But that is just about the sum total of knowledge, which is particularly frustrating since Venus is very much the same size as the Earth, so that detailed studies of the differences between the two planets potentially offer a most promising insight into how planets form and evolve.

One key piece of information has just become available, but has not yet been fully assimilated into our understanding of the planets. I refer, of course, to the pictures from the surface of Venus sent back by two Soviet space probes which soft-landed late in 1975. These show rocky terrain at both landing sites, but with important differences which hint at the changes now going on on Venus. One area has jumbled rocks with sharp edges – clear evidence that some form of geological activity occurs on Venus today since these are 'young' boulders in the sense that they must have been broken relatively recently from larger rocks. At the other site, a similar jumble of boulders shows much softer, rounded surfaces which must be the product of erosion. With no water on Venus, it is not obvious what the agents of this erosion are, but there are plenty of active ingredients in the atmosphere as possible candidates. And, of course, if one set of boulders is seen to be eroded then we know for sure that the other set is young, since we can see that erosion definitely does occur on Venus. So the two probes nicely complement each other. The veil of Venus has lifted slightly to reveal a tantalizing glimpse of geological – perhaps even geophysical – activity; but as yet we can still only guess that the structure of the planet is like that of the Earth.

If the structure of the interior of Venus is similar to that of the Earth – which it should be since they formed in the same part of the Solar System from, as far as we know, the same cloud of proto-planetary material – the absence of a magnetic field can be explained since the planet rotates only once every 243 Earth days, too slowly for the dynamo effect to operate in the way it does on Earth. Indeed, the rotation of the solid planet is much slower than the rotation of the clouds of the upper atmosphere of Venus; this atmospheric circulation sweeps around the planet every six Earth days, producing very strong winds. This violent atmosphere is the only part of Venus we know much about – but even here the planet poses more questions than answers, and as yet we do not even know just how these enormous circulating winds – perhaps up to 100 metres a second – are generated.

The atmosphere of Venus is a far more prominent feature than our own: the mass of the planet is very nearly the same as that of the Earth, yet the mass of the atmosphere is 100 times greater than that of our own atmosphere, although one might have thought that it would be hard for a planet closer to the Sun to keep even as much atmosphere as the Earth does. This massive atmosphere is chiefly made up of carbon dioxide, producing a strong greenhouse effect and heating the surface of the planet to more than 450°C. The reason for this is probably that Venus lies close enough to the Sun for the surface to be too hot for seas of water to form even before the greenhouse effect became important. On Earth, there is as much carbon dioxide as there is on Venus – but most of it is locked up in compounds in the surface rocks. Since even the primitive Venus atmosphere would have produced a strong greenhouse effect through the presence of all the water vapour outgassed from the surface but unable to condense as oceans, the rocky surface of the planet must have become so hot that all the carbon dioxide was driven out into the atmosphere. Meanwhile, and over the long period since then, the original water has been lost through other reactions, some of which have produced the sulphuric acid droplets which recent observations have shown high in the clouds of Venus.

Even so, it is far from easy to explain how Venus could have dried out so completely, and some theorists argue that it was always dry, with the outgassing of carbon dioxide alone sufficient to start off a greenhouse effect which then built up in a positive feedback. The problem then is, since Earth and Venus formed so close together, why should one planet have oceans of water and the other none at all?

There is a prospective test to find which school of thought is correct. If Venus initially possessed water vapour which has been dissociated by the effect of solar ultraviolet radiation, most of its hydrogen will have been lost to space, leaving a trace in the atmosphere rich in the heavier, slower-moving isotope deuterium. On the other hand, if Venus had no water to start with then the trace of hydrogen in its atmosphere today must have been trapped from the solar wind – which contains very little deuterium. So, in principle, all we need is a sample of the atmosphere of Venus to measure the proportion of light and heavy hydrogen in order to tell if the planet has always been dry, or whether it started out like the Earth and differs only through being closer to the heat of the Sun. It is still very much a puzzle why two planets of similar size formed in only slightly different parts of the Solar System should have such different atmospheres. However, the U.S. space programme includes a combined Venus orbiter/landing project for launch in 1978, and the workings of the hot, dry and windy atmosphere may be better understood when it becomes possible to drop instrumented balloon packages to float at different pressure levels through it. Most probably, the evolution of the Venus atmosphere has followed a path which could easily have been taken by our own atmosphere, and finding out why their two paths diverged will clearly tell us a great deal about the atmosphere of the Earth. As yet, however, the study of Venus has proved puzzling and a little disappointing for both students of the solid planet and atmospheric scientists.

If you are interested in living planets, our Moon is also something of a disappointment. To astronomers and planetary physicists, however, the Moon makes a valuable study just because it is so lifeless; with no atmosphere and no erosion, and with geologically inactive outer layers, the Moon preserves on its surface a record of the ancient history of the Solar System, and the rocks brought back by Apollo missions help to fill in the picture of events in the epoch just after the planets had formed. The age of this surface material is between 3,000 and 4,500 million years, and the oldest meteorites also formed about 4,500 million years ago, tying in with other evidence about the age of the Earth and Solar System (see Chapter One). But if you are looking for a geophysically active planet where physical processes still going on today can be compared with those going on here at home, the next planet out from the Earth-Moon system, Mars, amply makes up for the disappointments of both Venus and our Moon. Now well studied by a clutch of space probes, the most important prior to the Viking landings of 1976 being the orbiter Mariner 9, Mars provides a veritable bonanza of information to sift and interpret. The interpretations show that some kinds of geological (and perhaps geophysical) activity found on Earth also take place on Mars, and that at one time early in its history conditions there must have been rather like those on Earth at the time life originated.

Mariner 9 went into orbit around Mars in 1971 and, over a period of many months, produced a complete photographic map of the planet. This shows the planet as geologically active, possessing among other features the biggest known volcano in the Solar System, the Olympus Mons, which is as big as all the volcanoes of the Hawaiian chain put together and stands twenty-five kilometres above the surrounding countryside (Figure 16). There is also a rift valley system as extensive as the great rift valley system of East Africa, where although older faults are distinguished by the extent of erosion younger faults stand out clearly and show that activity continues on the planet.

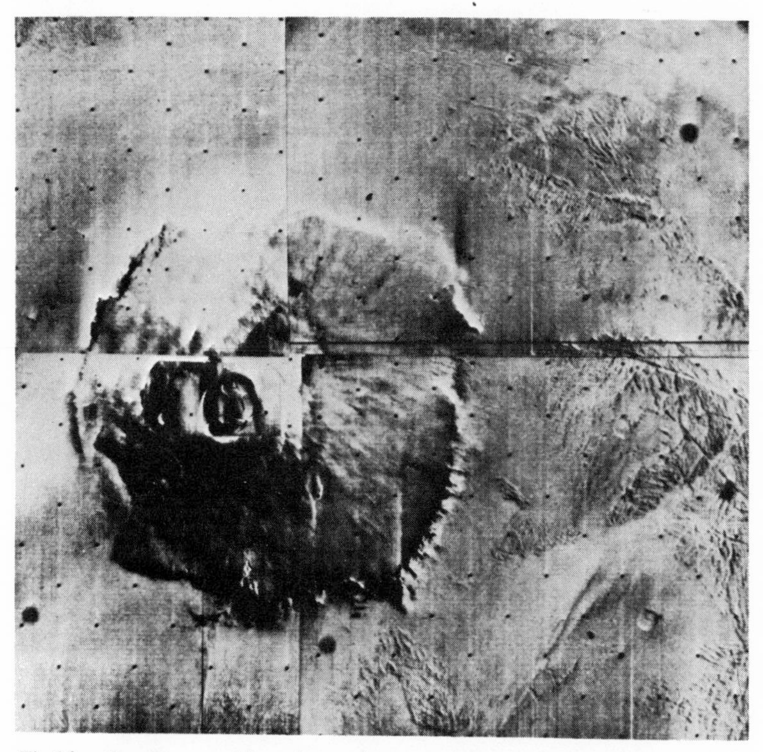

Fig.16 The Olympus Mons, largest known volcano in the Solar System.(NASA picture)

The northern hemisphere of Mars is low lying and relatively smooth, but the planet overall is pear-shaped with the southern hemisphere rising three kilometres above the mean radius, being more rugged and heavily cratered (implying greater age) and including the large volcanoes which lie across the equatorial regions. In very simplistic terms, this differentiation may correspond to the difference between continental and oceanic crust on Earth, with the low-lying larval plains of one hemisphere having formed over geological time as lighter molten rock from the interior has reached the surface, while on the 'continent' the upwelling material burst out through volcanic structures. No one has yet identified the equivalent of spreading ridges and deep trenches on Mars, and they may not exist, but even Harold Masursky, leader of the Mariner 9 TV imaging team and senior scientist at the U.S. Geological Survey's Center for Astrogeology, has speculated, 'Does the present surface of Mars resemble the Earth when "Pangea", the universal continent, existed before it was broken up by continental drift? Why are the Mars continental highlands restricted to one hemisphere? Was it by mantle convection? Can we learn about the early stages in the Earth's history by continued study of Mars?'*

Picking just one piece of recent geophysical speculation about Mars out of the hat, it may be that the enormous size of some martian volcanoes, especially the Olympus Mons, arises because although the planet is active it is not at the present time a site of continental drift. If the Olympus Mons is as big as all the Hawaiian chain put together, and if that chain has been produced by a hot spot punching successive holes through the moving crust above, could it not be that a similar hot spot on Mars has poured all of its productivity out into just one volcano through a single hole in a stationary piece of crust? To answer questions like that, we need more landers with rather more sophisticated geophysical equipment than the Viking probes – indeed, we probably need manned exploration of Mars. However, the martian atmosphere and climate may have been largely

* *New Science in the Solar System* (I.P.C. Magazines, London, 1975), p.38.

explained already, on the basis of Mariner 9 evidence and some inspired theorizing back on Earth.

As well as the direct evidence of geological activity, the Mariner 9 pictures show indirect evidence of former erosion from what seems to have been running water. Sinuous channels carved in the martian surface, apparent dried-up river deltas and many other signs of downhill flow by a fluid with properties like water can all be found – and although free water cannot exist in liquid form on Mars today, it is easier to account for these features in terms of flowing water under slightly different conditions in the past than to find another liquid which could leave the same traces but still not be around today.

What kind of change would be needed to allow water to flow on Mars? Today, the surface of the planet is very cold, and the atmospheric pressure there is very low. A bucket of water dumped there would quickly either freeze or evaporate – but it could no more remain liquid than carbon dioxide can exist as a liquid on Earth today. In order to have once had liquid water on Mars the pressure must have been higher – to stop evaporation – and the temperature must have been higher as well – to stop the water freezing. In fact, an increase in either pressure *or* temperature would do the trick, since if one changes the other follows suit, thanks to the greenhouse effect.

One way in which a rise in temperature might be produced comes from the pronounced wobble of Mars; every 100,000 years or so the polar caps 'nod' towards the Sun, and if this led to the release of frozen carbon dioxide locked up in them at present the whole planet could end up with a warmer, thicker blanket of atmosphere. The snag with this idea is that the dried-up martian rivers are much more than 100,000 years old – at least 500 *million* years old, and perhaps a couple of thousand million years old. This sobering insight into the time for which Mars has been dry comes from analysis of the distribution and numbers of the craters on Mars.

These craters, like those of Mercury and our Moon (and of Venus and the Earth), were produced by the impact of meteorites upon the planet, especially during the early stages of the development of the Solar System. The largest craters are undoubtedly the oldest, by and large, since they date from the time when many large fragments of rubble remained in the Solar System, leftovers from the planet-building process. And although many of the larger craters show signs of erosion, meaning that water flowed *after* they formed, the smaller, younger craters are still sharp-edged. In addition, there are a significant number of cases where craters overlie and obliterate parts of the presumed former river systems. However, a river is a pretty small target, and the number of meteorites hitting the martian surface even in a million years is not huge. So the rivers must have been around for quite some time in order to have been hammered in this way by the bombardment from space.

So the flowering of Mars must have been a very short-lived event early in the planet's youth. And this is the clue which has led scientists such as James Pollack, of the Ames Research Center, to produce what seems the definitive model of how Mars first gained – then lost – a warm, wet and thick blanket of atmosphere. The evidence of geological activity certainly makes it quite permissible to add the only other ingredient this model needs – outgassing of material early in the planet's life, to produce an atmosphere initially very similar to the initial atmosphere of the Earth, a reducing atmosphere rich in methane and ammonia, and paradoxically quite poisonous to the higher forms of life which exist on the Earth today.

If Mars did form this kind of thick, reducing atmosphere, similar to that of the early Earth, it would have evolved differently from our own atmosphere because the lighter gravitational pull of Mars would allow many atoms and some molecules to escape into space. Even at temperatures we regard as comfortable room heat, lighter gases – especially hydrogen – would be lost rapidly from the martian exosphere. So although an early, strong greenhouse effect would keep the planet warm initially, this very warmth would encourage and hasten the thinning of the atmosphere. After the river channels and water-carved features had formed, the planet cooled so much as the atmosphere thinned that the water froze into the polar caps and below the surface of the ground; with water vapour removed from the atmosphere, hydrogen lost to space and other compounds combined with surface rocks, this would leave the tenuous carbon dioxide atmosphere of the present-day Mars. But suppose that the warmer, wetter conditions lasted long enough for life to make the first steps on the evolutionary trail; could those life-forms have adapted and their descendants survived to the present day?

The answer may well be 'yes'. The amount of water locked up in the martian polar caps is probably enough to cover the entire planet to a depth of ten metres, and there may be as much water again in a layer of parafrost below the surface of the planet. Hardy terrestrial plants such as lichens can survive in very harsh environments, including dry valleys in Antarctica, and indeed some of these plants would probably survive on Mars if they could be transplanted. As for adapting to and evolving in a changing atmosphere, in many ways the changes in the Earth's atmosphere have been even more dramatic than the changes on Mars, and although oxygen is essential to everyone reading this book it would not strain the imagination too much to envisage some other form of intelligent life, approaching the Solar System from outside, deciding that life could not possibly survive in such a poisonous, corrosive atmosphere and that the early terrestrial life-forms must have all choked in their own wastes – especially oxygen!

The great advantage of oxygen for life that has adapted to its presence is that it reacts so energetically with other chemicals that it provides a good basis for powering an energetic life-form. You don't see very many active trees, or energetic lichens, both forms of life which stick to less dangerous ways of obtaining their energy. But since plants have lower energy requirements than animals such as man they can adapt to harsher environments. So by 1975 many planetologists were convinced that life could have started on Mars in the same way that life started on Earth, and that if it had started there was no reason to expect that some forms should not survive today.

The first quick look at the martian surface provided by the Viking landers can hardly be representative of such a large planet, but even the first Viking lander has given planetary scientists, a great deal of new food for thought. Pictures from the orbiting part of the Viking craft confirmed the general view of Mars as an active planet, once covered in running water (at least in middle latitudes), which emerged as such a surprise from the Mariner 9 mission. But the lander produced new surprises of its own.

Early results from the experiments on board the lander showed that something in the martian soil is very active. Moistened soil incubated in an Earth-like atmosphere inside the lander gave off a surprising amount of oxygen – far more, indeed, than soil from terrestrial deserts treated in the same way. This produced something of a three-day wonder at the end of July 1976, with some commentators jumping to the conclusion that living organisms might be giving off the oxygen. Furthermore, radioactive tracers added to the material in Viking's miniature, automated laboratory revealed carbon dioxide being produced for about seventy hours. If this was being produced by a chemical reaction, the flow should have been over in minutes; on the other hand, any biological process would be expected to continue for ten days or so. It seems, as of mid-August 1976, that there is something peculiar, in Earthly terms, about the martian soil, but no direct evidence has yet been found of life as we know it.

The presence of large amounts of oxygen bonded in the martian soil even by chemical means is, of course, intriguing in view of the possibility of a thicker, warmer blanket of air ever being produced again about the Red Planet. The hint of a new kind of process intermediate between biology and chemistry gives scientists from many disciplines something to ponder, and the prospect of monitoring martian winds and weather over a long period is clearly going to improve our understanding of planetary atmospheres, weather and climate. For the geophysicists, the big disappointment so far is the failure of Viking 1's seismometer. At the time of writing, hopes of monitoring martian 'earthquakes' are now pinned on the imminent landing of Viking 2 – but the success of Viking 1 alone is surely already enough to ensure that these will not be the last spacecraft from Earth to visit Mars.

Understandably, Mars is likely to remain a prime focus of attention for the planetary geophysicists for a long time to come. But our hypothetical visitor from beyond the Solar System, who decided that the Earth was an unlikely home for life, would probably be more interested in the gas giant Jupiter than in Mars or any of the terrestrial planets. Bigger than all the other planets put together, Jupiter not only still has the kind of reducing atmosphere in which we believe life originated on Earth, but is also kept warm because it is still radiating the heat left over from the formation of the planet. Warmth, methane and ammonia, and (in the cloud-tops at least) solar ultraviolet radiation should be ideal to start life developing, with the dense jovian atmosphere being analogous to the oceans on Earth. Perhaps, with no solid land to first crawl and then stand on, no intelligent life could develop, just as no intelligent life developed in our oceans. This may be dangerous parochialism from the viewpoint of one race on one small planet; a more certain handicap to the development of complicated life-forms on Jupiter is that the amount of solar radiation received is far weaker there, five times further from the Sun than we are. But there is certainly room for the science-fictioneers to speculate plausibly about life-forms floating through the jovian clouds like jellyfish through the sea.

Everything about Jupiter is big. The famous Great Red Spot is an impressive, but not dominating, feature of the planet – yet the spot is big enough to swallow up the Earth intact. So it is quite in keeping that the planet should possess a truly gigantic magnetosphere and a strong magnetic field, which succeeds in being both surprisingly like that of the Earth in some ways, and surprisingly different in others. To a space probe armed solely with magnetic detectors and a radio telescope, Earth and Jupiter would seem the most similar planets in the Solar System, and the only ones worth investigating in detail. Pioneer 10, the first probe from Earth to visit Jupiter, was armed with other detectors as well – but it was fully equipped to measure features of the jovian magnetosphere, as was its successor Pioneer 11.

Pioneer 10 moved into the jovian environment at a

distance of 108 jovian radii from the planet, when it crossed the bow shock wave (see Figure 17); at 96 planetary radii, it crossed the magnetopause and entered the magnetosphere. The magnetosphere of our own planet extends only to 13 Earth radii – just bigger than the radius of the planet Jupiter – but the jovian magnetic field seems most similar to the dipolar field of our own planet, except for its greater strength – in its middle regions – in terms of distance from the planet. Unlike the terrestrial magnetosphere, the outer regions of the jovian magnetosphere are very 'spongy', with the magnetopause being pushed in and out between 50 and 100 jovian radii as the solar wind varies. Beyond 20 jovian radii, the energetic particles of the jovian radiation belts stretch and distort the magnetic field as they push outwards. Between 20 and 3 radii, the field is dipolar and very similar to that of the Earth, but about eight times stronger. And within 3 radii of the planet the field is dominated by another component, a more complex quadrupole or octopole field, which has a very short range because its strength drops away with distance much more sharply than that of the dipole field. This complicated short-range structure superimposed on the dipole field is why the jovian magnetic field is both like and unlike that of our planet (Figure 17).

It seems very likely that the magnetism is produced by currents flowing in a liquid metallic core of the planet, and the complications may arise because of complications in these circulation patterns. Efforts to find out more about the detailed structure of the jovian magnetic field are therefore of particular importance since they provide a way of studying what goes on in the planet's interior. And since the liquid metallic core of the giant planet is believed to be made of hydrogen, in a form which cannot exist on Earth because of the great pressure needed to compress hydrogen into the liquid state, these observations will be of interest not just to planetary scientists but to the physicists who calculate how hydrogen should behave under extremes of pressure. But all these observations by future probes will be complicated, as were those of Pioneer 10 and Pioneer 11, by the disturbing effect of the four inner moons Amalthea, Io, Europa and Ganymede. These satellites all orbit *inside*

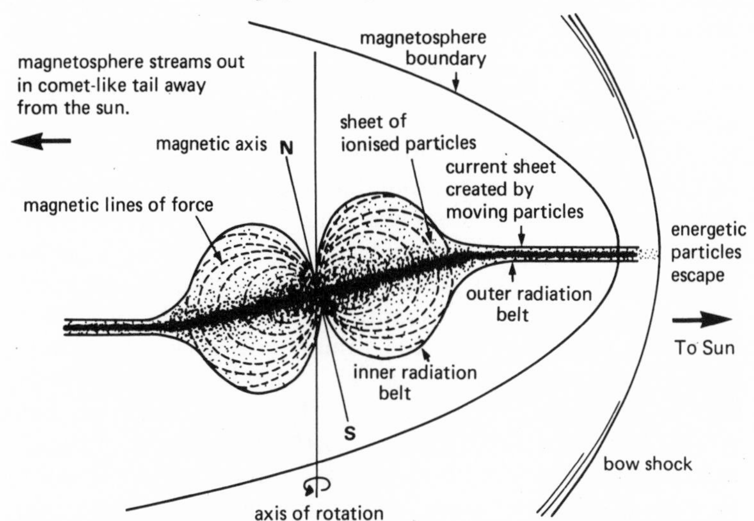

Fig.17 The magnetosphere of Jupiter, showing outer distorted region of the field and dipole dominated region; the scale does not make it possible to show the inner region of more complex fields. (From *New Science in the Solar System* [I.P.C. Magazines, London, 1975])

the magnetosphere in the equatorial plane, whereas our own Moon lies far outside the terrestrial magnetosphere; in the case of Io at least, this seems to have provided the moon with a tenuous atmosphere, constantly losing gas from the moon's weak gravity but constantly being replenished from the particles of the magnetosphere.

All of this is, perhaps, rather remote from the applications of geophysics and the new Earth science in general to finding out how best to use and ration the resources of our own planet. But few of the scientific benefits to civilization have come from strictly practical investigation of specific problems; most fundamental research is done through human curiosity, and only later put to use, and if the time ever comes that humankind ceases to be curious about the workings of the Universe it will probably be because humankind has ceased to exist. But in a way, with the immediate problems of overpopulation and lack of food and raw materials now facing us, this kind of insight into other parts of the Solar System does have some direct practical benefit. It has become a cliché of modern times that the most important direct benefit of the Apollo moon programme to mankind was that it brought home to millions the concept of 'Spaceship Earth', with its limited resources and too fertile population. But even a trip to the Moon hardly counts as putting humanity's collective toe into the cold of space. Even without looking half-way to the edge of our own Solar System we find in Jupiter a planet so large that our Earth could disappear into the jovian equivalent of a rather large tropical hurricane. In Chapter One we saw both how insignificant our whole Solar System is on the Universal scale of things, and how insignificant our civilization is compared with the history of one insignificant planet. In order to make good use of the new understanding of our changing Earth, we need not only the information and techniques now available but also the collective will to do something constructive with their aid. In scientific and technical terms, the problems can be solved, as I hope this book has shown. But is there sufficient political will for the necessary action? That depends on you.

Bibliography

Understanding the Earth, edited by I. G. Gass and P. J. Smith (Open University and M.I.T. Press, 2nd ed., 1972).
This is the best introduction to geophysics for anyone who is seriously interested in pursuing a career in the field. It contains contributions by the people actively involved in the 'revolution in the Earth sciences' and is thus a highly reliable record of great historical value. For this reason, Chapter Five of this book draws extensively from parts of *Understanding the Earth.*

Great World Atlas (Readers Digest, 2nd ed., 1968). Perhaps surprisingly, in view of its intended 'popular' market, this contains a lot of geological information and is more meaty than many coffee-table books.

Continental Drift, Ursula B. Marvin (Smithsonian Institution, Washington, 1973).
Historical account of the development of man's ideas about the changing face of the Earth.

Continental Drift, D. H. and M. P. Tarling (Pelican, London, 1974).
A cheap, handy, little pocket book, probably the best value of all the 'instant' guides to geophysics. Strongly recommended.

Restless Earth, Nigel Calder (B.B.C. Publications, London, 1972).
This book and the associated TV special have been the best products of the Nigel Calder/B.B.C. TV collaboration so far. The book has not dated because, of course, the basics of

plate tectonics were firmly established by 1972. If it has a flaw, it is the lack of any relation of the 'Restless Earth' to its place in the Solar System. The TV special is still repeated from time to time; be sure to see it if you get a chance.

The Origins of the Oceans and Continents, Alfred Wegener (Methuen, London, 1967).
A classic work.

The Jupiter Effect, John Gribbin and Stephen Plagemann (Macmillan, London and Walker, New York, 1974; paperback Vintage, New York, 1975 and Fontana, London, 1977). For more about the San Andreas Fault in particular and the relation of the whole Earth system to the rest of the Solar System in general.

Planet Earth; Readings from *Scientific American* (W. H. Freeman, Reading, 1975). See especially Chapters One to Three on Energy, including geothermal power, and Chapter Fifteen on Plate Tectonics and Mineral Resources.

Forecasts, Famines and Freezes, John Gribbin (Wildwood, London and Walker, New York, 1976). The climatic side of the Earth sciences.

New Science in the Solar System (I.P.C. Magazines, London, 1975). A *New Scientist* special review. Certainly the cheapest overview of the planets (at £1) and far from being the worst.

Earthquake Country, Robert Iacopi (Lane Books, California, 1973).
The best 'guide book' to the San Andreas Fault. Indispensable for the visitor to California.

The Making of the Scottish Landscape, R. N. Millman (Batsford, London, 1975).
Good description of how the rugged terrain left by former geological activity has moulded the development of a country.

Index